Wireless Networks

Series Editor

Xuemin Sherman Shen, University of Waterloo, Waterloo, ON, Canada

The purpose of Springer's Wireless Networks book series is to establish the state of the art and set the course for future research and development in wireless communication networks. The scope of this series includes not only all aspects of wireless networks (including cellular networks, WiFi, sensor networks, and vehicular networks), but related areas such as cloud computing and big data. The series serves as a central source of references for wireless networks research and development. It aims to publish thorough and cohesive overviews on specific topics in wireless networks, as well as works that are larger in scope than survey articles and that contain more detailed background information. The series also provides coverage of advanced and timely topics worthy of monographs, contributed volumes, textbooks and handbooks.

** Indexing: Wireless Networks is indexed in EBSCO databases and DPLB **

Wei Yang Bryan Lim • Jer Shyuan Ng •
Zehui Xiong • Dusit Niyato • Chunyan Miao

Federated Learning Over Wireless Edge Networks

Wei Yang Bryan Lim
Alibaba-NTU Joint Research Institute
Singapore, Singapore

Jer Shyuan Ng
Alibaba-NTU Joint Research Institute
Singapore, Singapore

Zehui Xiong
Singapore University of Technology and Design
Singapore, Singapore

Dusit Niyato
School of Computer Science and Engineering
Nanyang Technological University
Singapore, Singapore

Chunyan Miao
School of Computer Science and Engineering
Nanyang Technological University
Singapore, Singapore

ISSN 2366-1186 ISSN 2366-1445 (electronic)
Wireless Networks
ISBN 978-3-031-07837-8 ISBN 978-3-031-07838-5 (eBook)
https://doi.org/10.1007/978-3-031-07838-5

This Springer imprint is published by the registered company Springer Nature Switzerland AG
The registered company address is: Gewerbestrasse 11, 6330 Cham, Switzerland

Preface

The confluence of edge computing and artificial intelligence (AI) has driven the rise of edge intelligence, which leverages the storage, communication, and computation capabilities of end devices and edge servers to empower AI implementation at scale closer to where data is generated. An enabling technology of edge intelligence is the privacy-preserving machine learning paradigm known as federated learning (FL). Amid the increasingly stringent privacy regulations, FL will enable the development of applications that have to be built using sensitive user data and will continue to revolutionize service delivery in finance, Internet of Things (IoT), healthcare, and transport industries, among others. However, the implementation of FL is envisioned to involve thousands of heterogeneous distributed end devices that differ in terms of communication and computation resources, as well as the levels of willingness to participate in the collaborative model training process. The potential node failures, device dropouts, and stragglers effect are key bottlenecks that impede the effective, sustainable, and scalable implementation of FL.

In Chap. 1, we will first present a tutorial and survey on FL and highlight its role in enabling edge intelligence. This tutorial and survey provide readers with a comprehensive introduction to the forefront challenges and state-of-the-art approaches towards implementing FL at the edge. In consideration of resource heterogeneity at the edge networks, we then provide multifaceted solutions formulated via the interdisciplinary interplay of concepts derived from network economics, optimization, game theory, and machine learning towards improving the efficiency of resource allocation for implementing FL at scale amid information asymmetry. In Chap. 2, we devise a multi-dimensional contract-matching approach for optimized resource allocation for federated sensing and learning amid multi-dimensional sources of heterogeneities. In Chap. 3, we propose a joint-auction coalition formation framework towards facilitating communication-efficient FL networks aided by unmanned aerial vehicles (UAVs). In Chap. 4, we propose a two-level evolutionary game theoretic and auction approach to allocate and price resources to facilitate efficient edge intelligence. In Chap. 5, we recap the key points and discuss the promising research directions for future works.

We sincerely thank our collaborators for their contributions to the presented research works. Special thanks also go to the staff at Springer Nature for their help throughout the publication preparation process. Finally, we would like to take the chance to dedicate this book to celebrate the birth of Lim Chen Huan Theodore, son of Dr. Lim Wei Yang Bryan and Foo Feng Lin.

Singapore, Singapore — Wei Yang Bryan Lim
Singapore, Singapore — Jer Shyuan Ng
Singapore, Singapore — Zehui Xiong
Singapore, Singapore — Dusit Niyato
Singapore, Singapore — Chunyan Miao

Contents

1 Federated Learning at Mobile Edge Networks: A Tutorial 1
1.1 Introduction 1
1.2 Background and Fundamentals of Federated Learning 6
1.2.1 Federated Learning 6
1.2.2 Statistical Challenges of FL 8
1.2.3 FL Protocols and Frameworks 11
1.2.4 Unique Characteristics and Issues of FL 12
1.3 Communication Cost 12
1.3.1 Edge and End Computation 13
1.3.2 Model Compression 16
1.3.3 Importance-Based Updating 17
1.4 Resource Allocation 18
1.4.1 Worker Selection 21
1.4.2 Joint Radio and Computation Resource Management 25
1.4.3 Adaptive Aggregation 27
1.4.4 Incentive Mechanism 28
1.5 Privacy and Security Issues 32
1.5.1 Privacy Issues 33
1.5.2 Security Issues 37
1.6 Applications of Federated Learning for Mobile Edge Computing 39
1.6.1 Cyberattack Detection 42
1.6.2 Edge Caching and Computation Offloading 44
1.6.3 Base Station Association 47
1.6.4 Vehicular Networks 48
1.7 Conclusion and Chapter Discussion 50

2 Multi-dimensional Contract Matching Design for Federated Learning in UAV Networks 53
2.1 Introduction 53
2.2 System Model and Problem Formulation 56
2.2.1 UAV Sensing Model 58
2.2.2 UAV Computation Model 59

2.2.3 UAV Transmission Model 60
2.2.4 UAV and Model Owner Utility Modeling 61
2.3 Multi-dimensional Contract Design 61
2.3.1 Contract Condition Analysis 62
2.3.2 Conversion into a Single-Dimensional Contract 63
2.3.3 Conditions for Contract Feasibility 64
2.3.4 Contract Optimality 68
2.4 UAV-Subregion Assignment 70
2.4.1 Matching Rules 71
2.4.2 Matching Implementation and Algorithm 72
2.5 Performance Evaluation 73
2.5.1 Contract Optimality 73
2.5.2 UAV-Subregion Preference Analysis 76
2.5.3 Matching-Based UAV-Subregion Assignment 78
2.6 Conclusion and Chapter Discussion 80

3 Joint Auction–Coalition Formation Framework for UAV-Assisted Communication-Efficient Federated Learning 83
3.1 Introduction 83
3.2 System Model 86
3.2.1 Worker Selection 88
3.2.2 UAV Energy Model 90
3.3 Coalitions of UAVs 93
3.3.1 Coalition Game Formulation 94
3.3.2 Coalition Formation Algorithm 97
3.4 Auction Design 98
3.4.1 Buyers' Bids 99
3.4.2 Sellers' Problem 100
3.4.3 Analysis of the Auction 102
3.4.4 Complexity of the Joint Auction–Coalition Algorithm 104
3.5 Simulation Results and Analysis 105
3.5.1 Communication Efficiency in FL Network 106
3.5.2 Preference of Cells of Workers 107
3.5.3 Profit-Maximizing Behavior of UAVs 109
3.5.4 Allocation of UAVs to Cells of Workers 111
3.5.5 Comparison with Existing Schemes 114
3.6 Conclusion and Chapter Discussion 115

4 Evolutionary Edge Association and Auction in Hierarchical Federated Learning 117
4.1 Introduction 117
4.2 System Model and Problem Formulation 120
4.2.1 System Model 120
4.2.2 Lower-Level Evolutionary Game 121
4.2.3 Upper-Level Deep Learning Based Auction 121
4.3 Lower-Level Evolutionary Game 122
4.3.1 Evolutionary Game Formulation 122

4.3.2 Worker Utility and Replicator Dynamics 123
4.3.3 Existence, Uniqueness, and Stability of the Evolutionary Equilibrium 125
4.4 Deep Learning Based Auction for Valuation of Cluster Head 128
4.4.1 Auction Formulation... 128
4.4.2 Deep Learning Based Auction for Valuation of Cluster Heads 130
4.4.3 Monotone Transform Functions................................ 132
4.4.4 Allocation Rule 133
4.4.5 Conditional Payment Rule 134
4.4.6 Neural Network Training....................................... 134
4.5 Performance Evaluation 135
4.5.1 Lower-Level Evolutionary Game 136
4.5.2 Upper-Level Deep Learning Based Auction.................. 140
4.6 Conclusion and Chapter Discussion.................................... 143

5 Conclusion and Future Works.. 147

References... 151

Index... 165

List of Figures

Fig. 1.1 Edge AI approach brings AI processing closer to where data are produced. In particular, FL allows training on devices where the data are produced 3
Fig. 1.2 General FL training process involving *N* workers 7
Fig. 1.3 Approaches to increase computation at edge and end devices include (**a**) increased computation at end devices, e.g., more passes over dataset before communication, (**b**) two-stream training with global model as a reference, and (**c**) intermediate edge server aggregation 14
Fig. 1.4 Worker selection under the FedCS and Hybrid-FL protocol 22
Fig. 1.5 A comparison between (**a**) BAA by over-the-air computation that reuses bandwidth (above) and (**b**) OFDMA (below) that uses only the allocated bandwidth 26
Fig. 1.6 A comparison between (**a**) synchronous and (**b**) asynchronous FL 27
Fig. 1.7 Workers with unknown resource constraints maximize their utility only if they choose the bundle that best reflects their constraints 30
Fig. 1.8 Selective parameter sharing model 35
Fig. 1.9 GAN attack on collaborative deep learning 36
Fig. 1.10 An illustration of (**a**) conventional FL and (**b**) the proposed BlockFL architectures 40
Fig. 1.11 FL based attack detection architecture for IoT edge networks 43
Fig. 1.12 FL based (**a**) caching and (**b**) computation offloading 45
Fig. 2.1 System model involving UAV-subregion contract matching 55
Fig. 2.2 UAV node coverage vs. auxiliary types 75
Fig. 2.3 Contract rewards vs. auxiliary types 75
Fig. 2.4 Contract items vs. UAV utilities 75

Fig. 2.5 The model owner profits vs. UAV auxiliary types ... 76
Fig. 2.6 The UAV utility for each subregion vs. types ... 77
Fig. 2.7 UAV matching for homogeneous subregions ... 78
Fig. 2.8 UAV matching for subregions with different data quantities and coverage area ... 79
Fig. 2.9 UAV matching where $J > N$ ... 79
Fig. 3.1 System model consists of the cloud server (FL model owner), the vehicles and RSUs (selected FL workers), and the UAVs ... 85
Fig. 3.2 Illustration of the joint auction–coalition formation framework ... 87
Fig. 3.3 Distributed FL Network with 3 cells and 6 UAVs ... 105
Fig. 3.4 Communication time needed by UAVs and IoV vehicles ... 107
Fig. 3.5 Maximum number of iterations under different energy capacities ... 110
Fig. 3.6 Illustration of merge-and-split mechanism and allocation of UAVs to cells of workers ... 111
Fig. 3.7 Total profit and number of coalitions vs cooperation cost ... 113
Fig. 3.8 Total profit and size of coalitions vs number of iterations ... 113
Fig. 3.9 Comparison with existing schemes ... 114
Fig. 4.1 An illustration of the hierarchical system model ... 118
Fig. 4.2 Neural network architecture for the optimal auction ... 130
Fig. 4.3 Monotone transform functions ... 133
Fig. 4.4 Phase plane of the replicator dynamics ... 137
Fig. 4.5 Evolutionary equilibrium of population states for cluster 1 ... 137
Fig. 4.6 Evolution of population utilities ... 138
Fig. 4.7 Evolution of population states for population 1 ... 138
Fig. 4.8 Evolution of population states for population 2 ... 138
Fig. 4.9 Evolution of population states for population 3 ... 138
Fig. 4.10 Evolutionary dynamics under different learning rates ... 139
Fig. 4.11 Data coverage vs. varying fixed rewards in cluster 1 ... 139
Fig. 4.12 Data coverage vs. varying congestion coefficient in cluster 1 ... 140
Fig. 4.13 Population states in cluster 3 vs. varying population data for population 1 ... 140
Fig. 4.14 Revenue of cluster head 1 under different distribution of model owners ... 141
Fig. 4.15 Revenue of cluster head 2 under different distribution of model owners ... 141
Fig. 4.16 Revenue of cluster head 3 under different distribution of model owners ... 141
Fig. 4.17 Revenue vs data coverage of cluster heads ... 142
Fig. 4.18 Revenue of cluster head 1 under different approximation qualities ... 143
Fig. 4.19 Revenue of cluster head 2 under different approximation qualities ... 143

Fig. 4.20 Revenue of cluster head 3 under different approximation qualities 143

List of Tables

Table 1.1 An overview of selected surveys in FL and MEC 5
Table 1.2 Loss functions of common ML models 8
Table 1.3 Approaches to communication cost reduction in FL 19
Table 1.4 Approaches to resource allocation in FL 31
Table 1.5 The accuracy and attack success rates for no-attack scenario and attacks with 1 and 2 sybils in an FL system with MNIST dataset [1] 37
Table 1.6 A summary of attacks and countermeasures in FL 41
Table 1.7 FL based approaches for mobile edge network optimization 42
Table 2.1 Table of commonly used notations 56
Table 2.2 Table of key simulation parameters................................. 74
Table 2.3 UAV types for preference analysis................................... 77
Table 2.4 UAV type and preference for subregions 78
Table 3.1 Table of commonly used notations 89
Table 3.2 Simulation parameters 106
Table 3.3 Preference of cells for different coalitions 107
Table 3.4 Preference of UAVs for different cells.............................. 108
Table 3.5 Preference of cells for individual UAVs (bolded row implies top preference)... 108
Table 3.6 Revenue, cost of profit for different coalitions in each cell......... 110
Table 4.1 Simulation parameters 135

Chapter 1
Federated Learning at Mobile Edge Networks: A Tutorial

1.1 Introduction

Currently, there are nearly 7 billion connected Internet of Things (IoT) devices and 3 billion smartphones around the world [2]. These devices are equipped with increasingly advanced sensors, computing, and communication capabilities. As such, they can potentially be deployed for various crowdsensing tasks, e.g., for medical purposes [3] and air quality monitoring [4]. Coupled with the rise of Deep Learning (DL) [5], the wealth of data collected by end devices opens up countless possibilities for meaningful research and applications.

In the traditional cloud-centric approach, data collected by mobile devices are uploaded and processed centrally in a cloud-based server or data center. In particular, data collected by IoT devices and smartphones such as measurements [6], photos [7], videos [8], and location information [9] are aggregated at the data center [10]. Thereafter, the data are used to provide insights or produce effective inference models. However, this approach is no longer sustainable for the following reasons. Firstly, data owners are increasingly privacy sensitive. Following privacy concerns among consumers due to high profile cases of data leaks and data misuse, policy makers have responded with the implementation of data privacy legislation such as the European Commission's General Data Protection Regulation (GDPR) [11] and Consumer Privacy Bill of Rights in the USA [12]. In particular, the consent (GDPR Article 6) and data minimalization principle (GDPR Article 5) limit data collection and storage only to what is consumer-consented and absolutely necessary for processing. Secondly, a cloud-centric approach involves long propagation delays and incurs unacceptable latency [13] for applications in which real-time decisions have to be made, e.g., in self-driving car systems [14]. Thirdly, the transfer of

W. Y. B. Lim et al., *Federated Learning Over Wireless Edge Networks*, Wireless Networks, https://doi.org/10.1007/978-3-031-07838-5_1

raw data to the cloud for processing burdens the backbone networks.[1] This is especially so in tasks involving unstructured data, e.g., in video analytics [15]. This is exacerbated by the fact that cloud-centric training is relatively reliant on wireless communications [16]. As a result, this can potentially impede the development of new technologies.

With data sources mainly located outside the cloud today, Mobile Edge Computing (MEC) has naturally been proposed as a solution. In MEC, the computing and storage capabilities [13] of end devices and edge servers are leveraged to bring model training closer to where data are produced [17]. As defined in [16], an end–edge–cloud computing network comprises (a) end devices, (b) edge nodes, and (c) cloud server. For model training in conventional MEC approaches, a collaborative paradigm has been proposed in which training data are first sent to the edge servers for model training up to lower-level DNN layers, before more computation intensive tasks are offloaded to the cloud [18], [19] (Fig. 1.1). However, this arrangement incurs significant communication costs and is unsuitable for applications that require persistent training [16]. In addition, computation offloading and data processing at edge servers still involve the transmission of potentially sensitive personal data. This can discourage privacy-sensitive consumers from taking part in model training or even violate increasingly stringent privacy laws [11]. Although various privacy preservation methods, e.g., Differential Privacy (DP) [20], have been proposed, a number of users are still not willing to expose their private data for fear that their data may be inspected by external servers. In the long run, this discourages the development of technologies as well as new applications.

To guarantee that training data remain on personal devices and to facilitate collaborative machine learning of complex models among distributed devices, a decentralized ML approach called Federated Learning (FL) is introduced in [21]. In FL, mobile devices[2] use their local data to cooperatively train an ML model required by an FL server. They then send the model updates, i.e., the model's weights, to the FL server for aggregation. The steps are repeated in multiple rounds until a desirable accuracy is achieved. This implies that FL can be an enabling technology for ML model training at mobile edge networks. As compared to conventional cloud-centric ML model training approaches, the implementation of FL for model training at mobile edge networks features the following advantages:

1. *Minimizing requirement of network bandwidth:* Less information is required to be transmitted to the cloud. For example, instead of sending the raw data over for processing, participating devices only send the updated model parameters for

[1] Note that in some cases, the communication cost of FL is not insignificant due to complexity of large models, whereas the relatively fewer data samples a worker possesses are less costly to transmit. As such, we discuss communication cost reduction as well in this chapter.

[2] We use the term mobile devices and mobile edge in this chapter given that many of the works reviewed have focused on how to implement FL on resource-constrained edge devices such as IoT. However, note that the insights from this chapter can be similarly applied on edge networks in general.

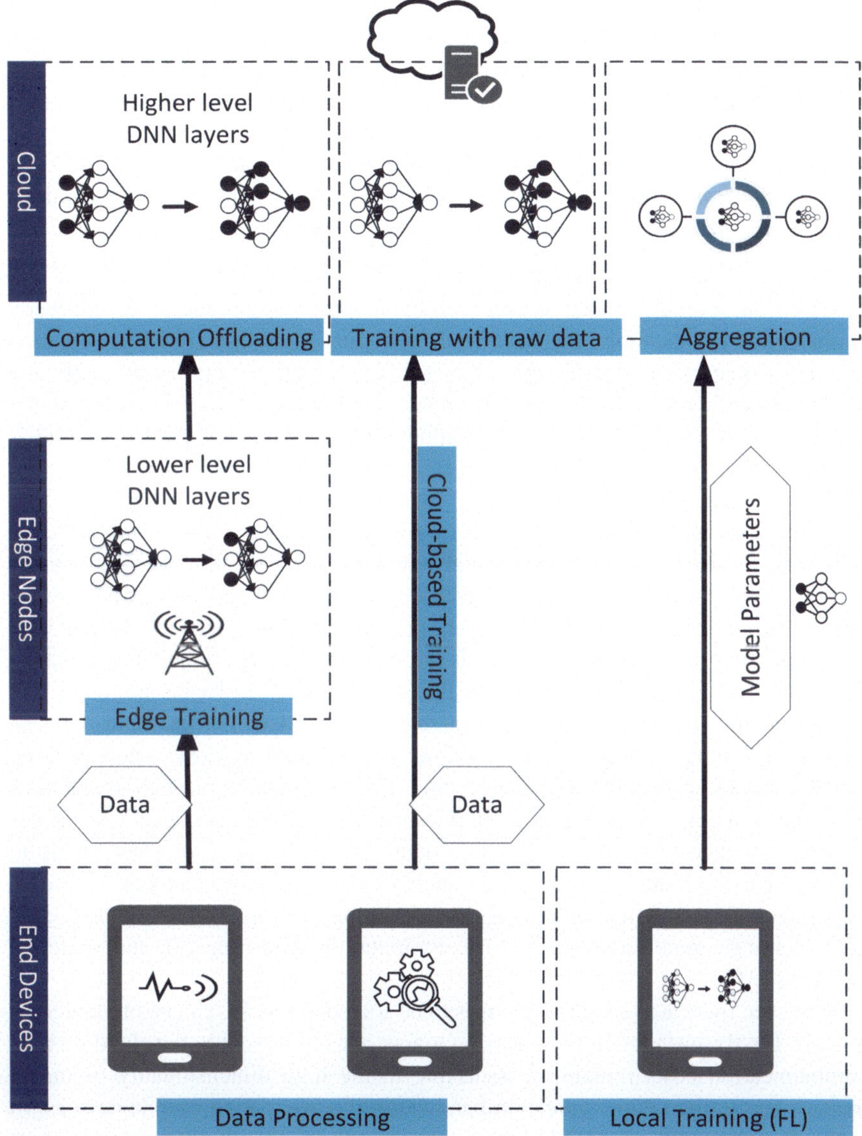

Fig. 1.1 Edge AI approach brings AI processing closer to where data are produced. In particular, FL allows training on devices where the data are produced

aggregation. As a result, this significantly reduces costs of data communication and relieves the burden on backbone networks.

2. *Privacy:* Following the above point, the raw data of users need not be sent to the cloud. Under the assumption that FL workers and servers are non-malicious, this

enhances user privacy and reduces the probability of eavesdropping to a certain extent. In fact, with enhanced privacy, more users will be willing to take part in collaborative model training, and so, better inference models can be built.
3. *Low latency:* With FL, ML models can be consistently trained and updated. Meanwhile, in the MEC paradigm, real-time decisions, e.g., event detection [22], can be made locally at the edge nodes or end devices. Therefore, the latency is much lower than when decisions are made in the cloud before transmitting them to the end devices. This is vital for time critical applications such as self-driving car systems in which the slightest delays can potentially be life threatening [14].

Given the aforementioned advantages, FL has seen recent successes in several applications. For example, the Federated Averaging algorithm (*FedAvg*) proposed in [21] has been applied to Google's Gboard [23] to improve next-word prediction models. In addition, several studies have also explored the use of FL in a number of scenarios in which data are sensitive in nature, e.g., to develop predictive models for diagnosis in health AI [24] and to foster collaboration across multiple hospitals [25] and government agencies [26].

Besides being an enabling technology for ML model training *at* mobile edge networks, FL has also been increasingly applied as an enabling technology *for* mobile edge network optimization. Given the computation and storage constraints of increasingly complex mobile edge networks, conventional network optimization approaches that are built on static models fare relatively poorly in modeling dynamic networks [16]. As such, a data-driven Deep Learning (DL) based approach [27] for optimizing resource allocation is increasingly popular. For example, DL can be used for representation learning of network conditions [28], whereas Deep Reinforcement Learning (DRL) can optimize decision making through interactions with the dynamic environment [29]. However, the aforementioned approaches require user data as an input and these data may be sensitive or inaccessible in nature due to regulatory constraints. As such, in this chapter, we also briefly discuss FL's potential to function as an enabling technology for optimizing wireless edge networks, e.g., in cell association [30], computation offloading [2], and vehicular networks [31].

However, there are several challenges to be solved before FL can be implemented at scale. Firstly, even though raw data no longer need to be sent to the cloud servers, communication costs remain an issue due to the high dimensionality of model updates and limited communication bandwidth of participating mobile devices. In particular, state-of-the-art DNN model training can involve the communication of millions of parameters for aggregation. Secondly, in a large and complex mobile edge network, the heterogeneity of participating devices in terms of data quality, computation power, and willingness to participate has to be well managed from the resource allocation perspective. Thirdly, FL does not guarantee privacy in the presence of malicious workers or aggregating servers. In particular, recent research works have clearly shown that a malicious worker may exist in FL and can infer the information of other workers just from the shared parameters alone. As such, privacy and security issues in FL still need to be considered.

Table 1.1 An overview of selected surveys in FL and MEC

Ref.	Subject	Contribution
[32]	FL	Introductory tutorial on categorization of FL architectures, e.g., vertical FL
[33]		FL in optimizing resource allocation for wireless networks while preserving data privacy
[34]		Tutorial on FL and discussions of implementation challenges in FL
[15]	MEC	Computation offloading strategy to optimize DL performance in edge computing
[35]		Survey on architectures and frameworks for edge intelligence
[36]		ML for IoT management, e.g., network management and security
[19]		Survey on computation offloading in MEC
[29]		Survey on DRL approaches to address issues in communications and networking
[37]		Survey on techniques for computation offloading
[38]		Survey on architectures and applications of MEC
[39]		Survey on computing, caching, and communication issues at mobile edge networks
[40]		Survey on the phases of caching and comparison among the different caching schemes
[13]		Survey on joint mobile computing and wireless communication resource management in MEC

In this chapter, we focus on the literature review and address the following topics:

1. We motivate the importance of FL as an important paradigm shift toward enabling collaborative ML model training. Then, we provide a concise tutorial on FL implementation and present to the reader a list of useful open-source frameworks that pave the way for future research on FL and the related applications.
2. We discuss the unique features of FL relative to a centralized ML approach and the resulting implementation challenges. For each of these challenges, we present to the reader a comprehensive discussion of existing solutions and approaches explored in the FL literature.
3. We discuss FL as an enabling technology for mobile edge network optimization. In particular, we discuss the current and potential applications of FL as a privacy-preserving approach for applications in edge computing.

Besides, we also consolidate the existing surveys on FL and mobile edge computing in Table 1.1.

The rest of this chapter is organized as follows. Section 1.2 introduces the background and fundamentals of FL. Section 1.3 reviews solutions provided to reduce communication costs. Section 1.4 discusses resource allocation approaches in FL. Section 1.5 discusses the security and privacy issues. Section 1.6 discusses the applications of FL for edge networks. Section 1.7 concludes the chapter.

1.2 Background and Fundamentals of Federated Learning

1.2.1 Federated Learning

Motivated by privacy concerns among data owners, the concept of FL is introduced in [21]. FL allows users to collaboratively train a shared model while keeping personal data on their devices, thus alleviating their privacy concerns. As such, FL can serve as an enabling technology for ML model training at mobile edge networks. For an introduction to the categorizations of different FL settings, e.g., vertical and horizontal FL, we refer the interested readers to [32].

In general, there are two main entities in the FL system: the data owners (i.e., *workers*) and the model owner (i.e., *FL server*). Let $\mathcal{N} = \{1, \ldots, N\}$ denote the set of N data owners, each of which has a private dataset $D_{i\in\mathcal{N}}$. Each data owner i uses its dataset D_i to train a *local model* $\mathbf{w}_i$ and send only the local model parameters to the FL server. Then, all collected local models are aggregated $\mathbf{w} = \cup_{i\in\mathcal{N}}\mathbf{w}_i$ to generate a *global model* $\mathbf{w}_G$. This is different from the traditional centralized training which uses $\mathbf{D} = \cup_{i\in\mathcal{N}} D_i$ to train a model $\mathbf{w}_T$, i.e., data from each individual source is aggregated first before model training takes place centrally.

A typical architecture and training process of an FL system is shown in Fig. 1.2. In this system, the data owners serve as the FL workers which collaboratively train an ML model required by an aggregate server. An underlying assumption is that the data owners are honest, which means they use their real private data to do the training and submit the true local models to the FL server. Of course, this assumption may not always be realistic [41] and we discuss the proposed solutions subsequently in Sect. 1.4.

In general, the FL training process includes the following three steps. Note that the *local* model refers to the model trained at each participating device, whereas the *global* model refers to the model aggregated by the FL server.

1. *Step 1 (Task initialization)*: The server decides the training task, i.e., the target application, and the corresponding data requirements. The server also specifies the hyperparameters of the global model and the training process, e.g., learning rate. Then, the server broadcasts the initialized global model $\mathbf{w}_G^0$ and task to selected workers.
2. *Step 2 (Local model training and update)*: Based on the global model $\mathbf{w}_G^t$, where t denotes the current iteration index, each worker, respectively, uses its local data and device to update the local model parameters $\mathbf{w}_i^t$. The goal of worker i in iteration t is to find optimal parameters $\mathbf{w}_i^t$ that minimize the loss function $L(\mathbf{w}_i^t)$, i.e.,

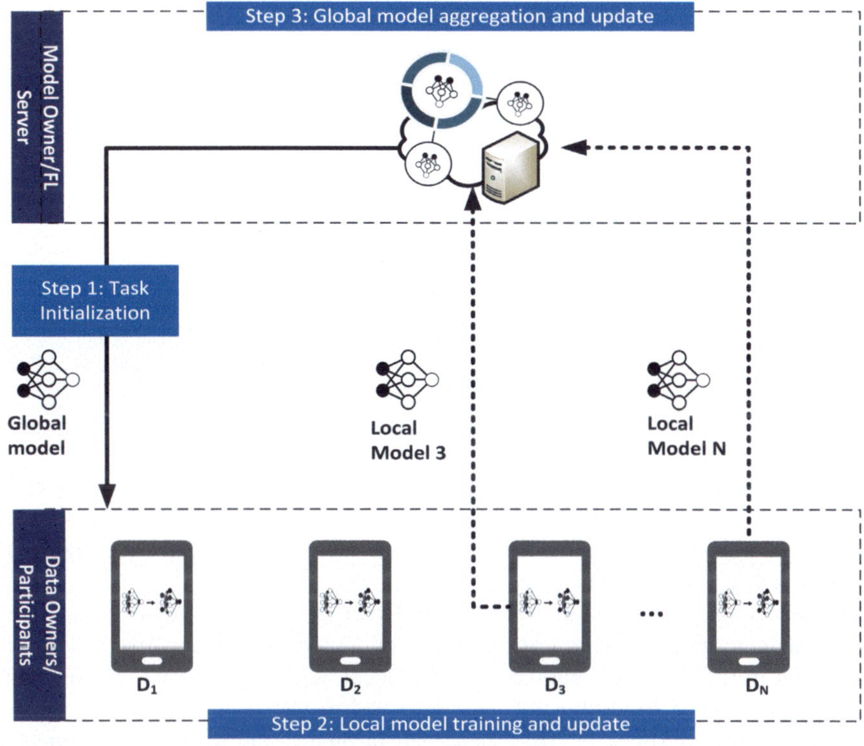

Fig. 1.2 General FL training process involving N workers

$$\mathbf{w}_i^{t^*} = \arg\min_{\mathbf{w}_i^t} L(\mathbf{w}_i^t). \tag{1.1}$$

The updated local model parameters are subsequently sent to the server.

3. *Step 3 (Global model aggregation and update)*: The server aggregates the local models from workers and then sends the updated global model parameters $\mathbf{w}_G^{t+1}$ back to the data owners.

   The server wants to minimize the global loss function $L(\mathbf{w}_G^t)$, i.e.,

$$L(\mathbf{w}_G^t) = \frac{1}{N}\sum_{i=1}^{N} L(\mathbf{w}_i^t). \tag{1.2}$$

Steps 2 and 3 are repeated until the global loss function converges or a desirable training accuracy is achieved.

Note that the FL training process can be used for different ML models that essentially use the SGD method such as Support Vector Machines (SVMs) [42], neural networks, and linear regression [43]. A training dataset usually contains a set

Table 1.2 Loss functions of common ML models

Model	Loss function $L(\mathbf{w}_i^t)$
Neural network	$\frac{1}{n}\sum_{j=1}^{n}(y_i - f(\mathbf{x}_j; \mathbf{w}))^2$ (Mean Squared Error)
Linear regression	$\frac{1}{2}\left\| y_j - \mathbf{w}^T\mathbf{x}_j \right\|^2$
K-means	$\sum_j \left\| \mathbf{x}_j - f(\mathbf{x}_j; \mathbf{w}) \right\|$ ($f(\mathbf{x}_j; \mathbf{w})$ is the centroid of all objects assigned to x_j's class)
Squared-SVM	$[\frac{1}{n}\sum_{j=1}^{n} \max(0, 1 - y_j(\mathbf{w}^T\mathbf{x}_j - bias))] + \lambda \left\| \mathbf{w}^T \right\|^2$ ($bias$ is the bias parameter and λ is const.)

of n data feature vectors $\mathbf{x} = \{\mathbf{x}_1, \ldots, \mathbf{x}_n\}$ and a set of corresponding data labels[3] $\mathbf{y} = \{y_1, \ldots, y_n\}$. In addition, let $\hat{y_j} = f(\mathbf{x}_j; \mathbf{w})$ denote the predicted result from the model $\mathbf{w}$ updated/trained by data vector x_j. Table 1.2 summarizes several loss functions of common ML models [44].

Global model aggregation is an integral part of FL. A straightforward and classical algorithm for aggregating the local models is the *FedAvg* algorithm proposed in [21], which is similar to that of local SGD [45]. The pseudocode for *FedAvg* is given in Algorithm 1.1. As described in Step 1 above, the server first initializes the task (lines 11–16). Thereafter, in Step 2, the worker i implements the local training and optimizes the target in (1.1) on minibatches from the original local dataset (lines 2–8). Note that a minibatch refers to a randomized subset of each worker's dataset. At the tth iteration (line 17), the server minimizes the global loss in (1.2) by the averaging aggregation which is formally defined as

$$\mathbf{w}_G^t = \frac{D_i}{\sum_{i \in \mathcal{N}} D_i} \sum_{i=1}^{N} \mathbf{w}_i^t. \tag{1.3}$$

The FL training process is iterated till the global loss function converges or a desirable accuracy is achieved.

1.2.2 Statistical Challenges of FL

In traditional distributed ML, the central server has access to the whole training dataset. As such, the server can split the dataset into subsets that follow similar

[3] In the case of unsupervised learning, there is no data label.

Algorithm 1.1 Federated averaging algorithm [21]

Require: Local minibatch size B, number of workers m per iteration, number of local epochs E, and learning rate η.
Ensure: Global model $\mathbf{w}_G$.

[Worker i]
LocalTraining(i, $\mathbf{w}$):
Split local dataset D_i to minibatches of size B which are included into the set $\mathcal{B}_i$.
for each local epoch j from 1 to E **do**
    for each $b \in \mathcal{B}_i$ **do**
        $\mathbf{w} \leftarrow \mathbf{w} - \eta \Delta L(\mathbf{w}; b)$ (η is the learning rate and ΔL is the gradient of L on b.)
    end for
end for
[Server]
Initialize $\mathbf{w}_G^0$
for each iteration t from 1 to T **do**
    Randomly choose a subset $\mathcal{S}_t$ of m workers from $\mathcal{N}$
    for each participant $i \in \mathcal{S}_t$ **parallelly do**
        $\mathbf{w}_i^{t+1} \leftarrow$ **LocalTraining**(i, $\mathbf{w}_G^t$)
    end for
    $\mathbf{w}_G^t = \frac{1}{\sum_{i \in \mathcal{N}} D_i} \sum_{i=1}^{N} D_i \mathbf{w}_i^t$ (Averaging aggregation)
end for

distributions. The subsets are subsequently sent to participating nodes for distributed training. However, this approach is impractical for FL since the local dataset is only accessible by the worker.

In the FL setting, the workers may have local datasets that follow different distributions, i.e., the datasets of workers are non-IID. While the authors in [21] show that the aforementioned *FedAvg* algorithm is able to achieve desired accuracy even when data are non-IID across workers, the authors in [46] found otherwise. For example, the accuracy of a *FedAvg*-trained CNN model has 51% lower accuracy than the centrally trained CNN model for CIFAR-10 [47]. This deterioration in accuracy is further shown to be quantified by the earth mover's distance (EMD) [48], i.e., the difference in FL worker's data distribution as compared to the population distribution. As such, when data are non-IID and highly skewed, data sharing is proposed in which a shared dataset with uniform distribution across all classes is sent by the FL server to each FL worker. Then, the worker trains its local model on its private data together with the received data. The simulation result shows that accuracy can be increased by 30% with 5% shared data due to reduced EMD. However, a common dataset may not always be available for sharing by the FL server. An alternative solution to gather contributions toward building the common dataset is subsequently discussed in Sect. 1.4.

The authors in [49] also find that global imbalance, i.e., the situation in which the collection of data held across all FL workers is class imbalanced, leads to a deterioration in model accuracy. As such, the Astraea framework is proposed. On initialization, the FL workers first send their data distribution to the FL server. A rebalancing step is introduced before training begins in which each worker performs data augmentation [50] on the minority classes, e.g., through random rotations and shifts. After training on the augmented data, a mediator is created to coordinate intermediate aggregation, i.e., before sending the updated parameters to the FL server for global aggregation. The mediator selects workers with data distributions that best contribute to a uniform distribution when aggregated. This is done through a greedy algorithm approach to minimize the Kullback–Leibler Divergence [51] between local data and uniform distribution. The simulation results show accuracy improvement when tested on imbalanced datasets.

Given the heterogeneity of data distribution across devices, there has been an increasing number of studies that borrow concepts from multi-task learning [52] to learn separate but structurally related models for each worker. Instead of minimizing the conventional loss function presented previously in Table 1.2 of Sect. 1.2.1 (Page 15), the loss function is modified to also model the relationship amongst tasks [53]. Then, the MOCHA algorithm is proposed in which an alternating optimization approach [54] is used to approximately solve the minimization problem. Interestingly, MOCHA can also be calibrated based on the resource constraints of a participating device. For example, the quality of approximation can be adaptively adjusted based on network conditions and CPU states of the participating devices. However, MOCHA cannot be applied to non-convex DL models.

Similarly, the authors in [55] also borrow concepts from multi-task learning to deal with the statistical heterogeneity in FL. The FEDPER approach is proposed in which all FL workers share a set of base layers trained using the *FedAvg* algorithm. Then, each worker separately trains another set of personalization layers using its own local data. In particular, this approach is suitable for building recommender's systems given the diverse preferences of workers. The authors show empirically using the Flickr-AES dataset [56] that the FEDPER approach outperforms a pure *FedAvg* approach since the personalization layer is able to represent the personal preference of an FL worker. However, it is worth noting that the collaborative training of the base layers is still important to achieve a high test accuracy, since each worker has insufficient local data samples for personalized model training.

Apart from data heterogeneity, the convergence of a distributed learning algorithm is always a concern. A higher convergence rate helps to save a large amount of time and resources for the FL workers and also significantly increases the success rate of the federated training since fewer communication rounds imply reduced worker dropouts. To ensure convergence, the study in [57] proposes *FedProx*, which modifies the loss function to also include a tunable parameter that restricts how much local updates can affect the prevailing model parameters. The *FedProx* algorithm can be adaptively tuned, e.g., when training loss is increasing, model updates can be tuned to affect the current parameters less. Similarly, the authors in [58] also propose the *LoAdaBoost FedAvg* algorithm to complement the aforementioned data sharing approach [46] in ML on medical data. In *LoAdaBoost FedAvg*, workers train the model on their local data and compare the cross-entropy loss with the median loss from the *previous* training round. If the current cross-entropy loss is higher, the model is retrained before global aggregation so as to increase learning efficiency. The simulation results show that faster convergence is achieved as a result.

In fact, the statistical challenges of FL coexist with other issues that we explore in subsequent sections. For example, the communication costs incurred in FL can be reduced by faster convergence. Similarly, resource allocation policies can also be designed to solve statistical heterogeneity. As such, we revisit these concepts in greater detail subsequently.

1.2.3 FL Protocols and Frameworks

The main open-source frameworks for FL that have been developed are as follows:

1. *TensorFlow Federated (TFF)*: TFF [59] is a framework based on Tensorflow developed by Google for decentralized ML and other distributed computations. TFF consists of two layers (a) FL and (b) Federated Core (FC). The FL layer is a high-level interface that allows the implementation of FL to existing TF models without the user having to apply the FL algorithms personally. The FC layer combines TF with communication operators to allow users to experiment with customized and newly designed FL algorithms.
2. *PySyft*: PySyft [60] is a framework based on PyTorch for performing encrypted, privacy-preserving DL and implementations of related techniques, such as Secure Multiparty Computation (SMPC) and DP, in untrusted environments while protecting data. Pysyft is developed such that it retains the native Torch interface, i.e., the ways to execute all tensor operations remain unchanged from that of PyTorch. When a SyftTensor is created, a LocalTensor is automatically created to also apply the input command to the native PyTorch tensor. To simulate FL, workers are created as *Virtual Workers*. Data, i.e., in the structure of tensors, can be split and distributed to the Virtual Workers as a simulation of a practical FL setting. Then, a PointerTensor is created to specify the data owner and storage location. In addition, model updates can be fetched from the Virtual Workers for global aggregation.
3. *LEAF*: An open-source framework [61] of datasets that can be used as benchmarks in FL, e.g., Federated Extended MNIST (FEMNIST), and Sentiment140 [62]. In these datasets, the writer or the user is assumed to be a worker in FL, and their corresponding data are taken to be the local data held in their personal devices. The implementation of newly designed algorithms on these benchmark datasets allows reliable comparison across studies.
4. *FATE*: Federated AI Technology Enabler (FATE) is an open-source framework by WeBank [63] that supports the federated and secure implementation of several ML models.

1.2.4 Unique Characteristics and Issues of FL

Besides the statistical challenges we have presented in Sect. 1.2.2, FL has some unique characteristics and features [64] as compared to other distributed ML approaches:

1. *Slow and unstable communication*: In the traditional distributed training in a data center, the communication environment can be assumed to be perfect where the information transmission rate is very high and there is no packet loss. However, these assumptions are not applicable to the FL environment where heterogeneous devices are involved in training. For example, the Internet upload speed is typically much slower than download speed [65]. Also, some workers with unstable wireless communication channels may consequently drop out due to disconnection from the Internet.
2. *Heterogeneous devices*: Apart from bandwidth constraints, FL involves heterogeneous devices with varying resource constraints. For example, the devices can have different computing capabilities, i.e., CPU states and battery levels. The devices can also have different levels of *willingness* to participate, i.e., FL training is resource consuming, and given the distributed nature of training across numerous devices, there is a possibility of free ridership.
3. *Privacy and security concerns*: As we have previously discussed, data owners are increasingly privacy sensitive. However, malicious workers are able to infer sensitive information from shared parameters, which potentially negates privacy preservation. In addition, we have previously assumed that all workers and FL servers are trustful.

These unique characteristics of FL lead to several practical issues in FL implementation mainly in three aspects that we now proceed to discuss, i.e., (a) statistical challenges (b) communication costs, and (c) resource allocation. In the following sections, we review related work that addresses each of these issues.

1.3 Communication Cost

In FL, a number of rounds of communications between the workers and the FL server may be required to achieve a target accuracy. For complex DL model training involving, e.g., CNN, each update may comprise millions of parameters [66]. The high dimensionality of the updates can result in the incurrence of high communication costs and can lead to a training bottleneck. In addition, the bottleneck can be worsened due to (a) unreliable network conditions of participating devices [67] and (b) the asymmetry in Internet connection speeds in which upload speed is faster than download speed, resulting in delays in model uploads by workers [65]. As such, there is a need to improve the communication efficiency of FL. The following approaches to reduce communication costs are considered:

1. *Edge and End Computation*: In the FL setup, the communication cost often dominates computation cost [21]. The reason is that on-device dataset is relatively small, whereas mobile devices of workers have increasingly fast processors. On the other hand, workers may be willing to participate in the model training only if they are connected to Wi-Fi [65]. As such, more computation can be performed on edge nodes or on end devices before each global aggregation so as to reduce the number of communication rounds needed for the model training. In addition, algorithms to ensure faster convergence can also reduce the number of communication rounds involved, at the expense of more computation on edge servers and end devices.
2. *Model Compression*: This is a technique commonly used in distributed learning [68]. Model or gradient compression involves the communication of an update that is transformed to be more compact, e.g., through sparsification, quantization, or subsampling [69], rather than the communication of full update. However, since the compression may introduce noise, the objective is to reduce the size of update transferred during each round of communication while maintaining the quality of trained models [70].
3. *Importance-Based Updating*: This strategy involves selective communication such that only the important or relevant updates [71] are transmitted in each communication round. In fact, besides saving on communication costs, omitting some updates from workers can even improve the global model performance.

1.3.1 Edge and End Computation

To decrease the number of communication rounds, the additional computation can be performed on participating end devices before each iteration of communication for global aggregation (Fig. 1.3a). The authors in [21] consider two ways to increase computation on participating devices: (a) increasing parallelism in which more workers are selected to participate in each round of training and (b) increasing computation per worker whereby each worker performs more local updates before communication for global aggregation. A comparison is conducted for the FederatedSGD (*FedSGD*) algorithm and the proposed *FedAvg* algorithm. For the *FedSGD* algorithm, all workers are involved and only one pass is made per training round in which the minibatch size comprises the entirety of the worker's dataset. This is similar to the full-batch training in centralized DL frameworks. For the proposed *FedAvg* algorithm, the hyperparameters are tuned such that more local computations are performed by the workers. For example, the worker can make more passes over its dataset or use a smaller local minibatch size to increase computation before each communication round. The simulation results show that increased parallelism does not lead to significant improvements in reducing communication cost once a certain threshold is reached. As such, more emphasis is placed on increasing computation per worker while keeping the fraction of selected workers constant. For MNIST

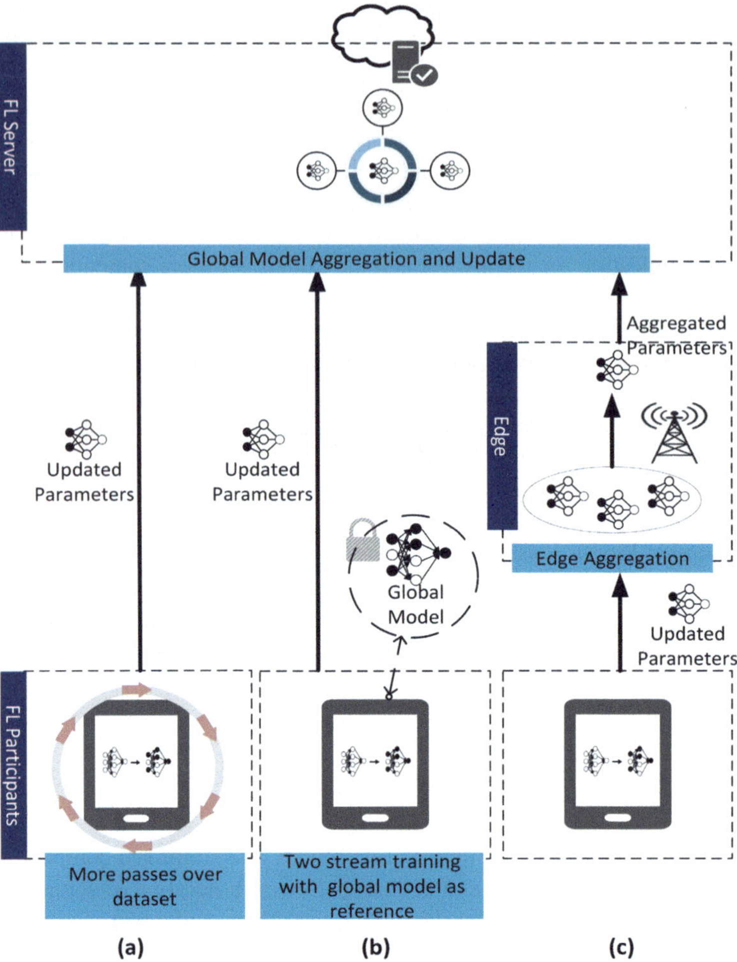

Fig. 1.3 Approaches to increase computation at edge and end devices include (**a**) increased computation at end devices, e.g., more passes over dataset before communication, (**b**) two-stream training with global model as a reference, and (**c**) intermediate edge server aggregation

CNN simulations, increased computation using the proposed *FedAvg* algorithm can reduce communication rounds when the dataset is IID. For non-IID dataset, the improvement is less significant using the same hyperparameters. For long short-term memory (LSTM) simulations [72], improvements are more significant even for non-IID data. However, as we will discuss later, the non-IID data distribution is a key concern in FL. The empirical observation of improvements found in [21] is not consistent when explored in other studies. Instead, more precise methods are proposed to deal with non-IID distributions.

As an extension, the authors in [73] also validate that a similar concept as that of [21] works for vertical FL. In vertical FL, collaborative model training is conducted across the same set of workers with different data features. The Federated Stochastic Block Coordinate Descent (FedBCD) algorithm is proposed in which each participating device performs multiple local updates first before communication for global aggregation. In addition, convergence guarantee is also provided with an approximate calibration of the number of local updates per interval of communication.

Another way to decrease communication cost can also be through modifying the training algorithm to increase convergence speed, e.g., through the aforementioned *LoAdaBoost FedAvg* in [58]. Similarly, the authors in [74] also propose increased computation on each participating device by adopting a two-stream model (Fig. 1.3b) commonly used in transfer learning and domain adaptation [75]. During each training round, the global model is received by the worker and fixed as a reference in the training process. During training, the worker learns not just from local data but also from other workers with reference to the fixed global model. This is done through the incorporation of Maximum Mean Discrepancy (MMD) into the loss function. MMD measures the distance between the means of two data distributions [75], [76]. By minimizing MMD loss between the local and global models, the worker can extract more generalized features from the global model, thus accelerating the convergence of the training process and reducing communication rounds. The simulation results on the CIFAR-10 and MNIST dataset using DL models such as AlexNet [77] and 2-CNN, respectively, show that the proposed two-stream FL can reach the desirable test accuracy in 20% fewer communication rounds even when data are non-IID. However, while convergence speed is increased, more computation resources have to be consumed by end devices for the aforementioned approaches. Given the energy constraints of participating mobile devices in particular, this necessitates resource allocation optimization that we subsequently discuss in Sect. 1.4.

While the aforementioned studies consider increasing computation on participating *devices*, the authors in [78] propose an edge computing inspired paradigm in which proximate edge *servers* can serve as intermediary parameter aggregators given that the propagation latency from worker to the edge server is smaller than that of the worker–cloud communication (Fig. 1.3c). A hierarchical FL (*HierFAVG*) algorithm is proposed whereby for every few local worker updates, the edge server aggregates the collected local models. After a predefined number of edge server aggregations, the edge server communicates with the cloud for global model aggregation. As such, the communication between the workers and the cloud occurs only once after an interval of multiple local updates. Comparatively, for the *FedAvg* algorithm proposed in [21], the global aggregation occurs more frequently since no intermediate edge server aggregation is involved. The authors further prove the convergence of *HierFAVG* for both convex and non-convex objective functions given non-IID user data. The simulation results show that for the same number of local updates between two global aggregations, more intermediate edge aggregations before each global aggregation can lead to reduced communication overhead as compared to the *FedAvg* algorithm. This result holds for both IID and non-IID data, implying that intermediate aggregation on edge servers may be implemented on top of *FedAvg* to reduce communication costs. However, when applied to non-IID data, the simulation results show that *HierFAVG* fails to converge to the desired accuracy level (90%) in some instances, e.g., when edge–cloud divergence is large or when there are many edge servers involved. As such, further study is required to better understand the tradeoffs between adjusting local and edge aggregation intervals, so as to ensure that the parameters of the *HierFAVG* algorithm can be optimally calibrated to suit other settings. Nevertheless, *HierFAVG* is a promising approach for the implementation of FL at mobile edge networks, since it leverages the proximity of intermediate edge servers to reduce communication costs and potentially relieve the burden on the remote cloud.

1.3.2 Model Compression

To reduce communication costs, the authors in [65] propose structured and sketched updates to reduce the size of model updates sent from workers to the FL server during each communication round.

Structured updates restrict worker updates to have a prespecified structure, i.e., low rank and random mask. For the low rank structure, each update is enforced to be a low rank matrix expressed as a product of two matrices. Here, one matrix is generated randomly and held constant during each communication round, whereas the other is optimized. As such, only the optimized matrix needs to be sent to the server. For the random mask structure, each worker update is restricted to be a sparse matrix following a predefined random sparsity pattern generated independently during each round. As such, only the nonzero entries are required to be sent to the server.

On the other hand, *sketched updates* refer to the approach of encoding the update in a compressed form before communication with the server, which subsequently decodes the updates before aggregation. One example of the sketched update is the subsampling approach, in which each worker communicates only a random subset of the update matrix. The server then averages the subsampled updates to derive an unbiased estimate of the true average.

The simulation results on the CIFAR-10 image classification task show that for structured updates, the random mask performs better than that of the low rank approach. The random mask approach also achieves higher accuracy than sketching approaches since the latter involves the removal of some information obtained during training. However, the combination of all three sketching tools, i.e., subsampling, quantization, and rotation, can achieve a higher compression rate and faster convergence, albeit with some sacrifices in accuracy. This suggests that for practical implementation of FL where there are many workers available, more workers can be selected for training per round so that subsampling can be more aggressive to reduce communication costs.

In addition to the subsampling and quantization approaches, the federated dropout approach is also considered in which a fixed number of activation functions at each fully connected layer are removed to derive a smaller sub-model. The sub-model is then sent to the workers for training. The updated sub-model can then be mapped back to the global model to derive a complete DNN model with all weights updated during subsequent aggregation. This approach reduces the server-to-worker communication cost and also the size of worker-to-server updates. In addition, local computation is reduced since fewer parameters have to be updated. The simulations are performed on MNIST, CIFAR-10, and EMNIST [79] datasets. For the lossy compression, it is shown that the subsampling approach taken by [65] does not reach an acceptable level of performance. The reason is that the update errors can be averaged out for worker-to-server uploads but not for server-to-worker downloads. On the other hand, quantization with Kashin's representation can achieve the same performance as the baseline without compression while having communication cost reduced by nearly 8 times when the model is quantized to 4 bits. For federated dropout approaches, the results show that a dropout rate of 25% of weight matrices of fully connected layers (or filters in the case of CNN) can achieve acceptable accuracy in most cases while ensuring around 43% reduction in the size of models communicated. However, if dropout rates are more aggressive, the convergence of the model can be slower.

The aforementioned two studies suggest useful model compression approaches in reducing communication costs for both server-to-worker and worker-to-server communications. As one may expect, the reduction in communication costs comes with sacrifices in model accuracy. It will thus be useful to formalize the compression–accuracy tradeoffs since this varies for different tasks.

1.3.3 Importance-Based Updating

Based on the observation that most parameter values of a DNN model are sparsely distributed and close to zero [80], the authors in [71] propose the edge Stochastic Gradient Descent (eSGD) algorithm that selects only a small fraction of important gradients to be communicated to the FL server for parameter update during each communication round. The eSGD algorithm keeps track of loss values at two consecutive training iterations. If the loss value of the current iteration is smaller than the preceding iteration, this implies that current training gradients and model parameters are important for training loss minimalization. As a result, their respective hidden weights are assigned a positive value. In addition, the gradient is also communicated to the server for parameter updates. Once this does not hold, i.e., the loss increases as compared to the previous iteration, other parameters are selected to be updated based on their hidden weight values. A parameter with a larger hidden weight value is more likely to be selected since it has been labeled as important many times during training. To account for small gradient values that can delay convergence if they are ignored and not updated completely [81], these gradient values are accumulated as residual values. Since the residuals may arise from different training iterations, each update to the residual is weighted with a discount factor using the momentum correction technique [82]. Once the accumulated residual gradient reaches a threshold, they are chosen to replace the least important gradient coordinates according to the hidden weight values. The simulation results show that eSGD with a 50% drop ratio can achieve higher accuracy than that of the threshold SGD algorithm proposed in [80], which uses a fixed threshold value to determine which gradient coordinates to drop. eSGD can also save a large proportion of gradient size communicated. However, eSGD still suffers from accuracy loss as compared to standard SGD approaches.

While [71] studies the selective communication of gradients, the authors in [67] propose the Communication-Mitigated Federated Learning (CMFL) algorithm that uploads only relevant local model updates to reduce communication costs while guaranteeing global convergence. In each iteration, a worker's local update is first compared with the global update to identify if the update is relevant. A relevance score is computed where the score equates to the percentage of the same sign parameters in the local and global update. In fact, the global update is not known in advance before aggregation. As such, the global update made in the previous iteration is used as an estimate for comparison since it was found empirically that more than 99% of the normalized differences between two sequential global updates are smaller than 0.05 in both MNIST CNN and Next-Word-Prediction LSTM. An update is considered to be irrelevant if its relevance score is smaller than a predefined threshold. The simulation results show that CMFL requires 3.47 times and 13.97 times fewer communication rounds to reach 80% accuracy for MNIST CNN and Next-Word-Prediction LSTM, respectively, as compared to the benchmark *FedAvg* algorithm. In addition, CMFL can save significantly more communication rounds as compared to Gaia. Note that Gaia is a geo-distributed ML approach suggested in [83] which measures relevance based on *magnitude* of updates rather than sign of parameters. When applied with the aforementioned MOCHA algorithm 1.2.2 [53], CMFL can reduce communication rounds by 5.7 times for the Human Activity Recognition dataset [84] and 3.3 times for the Semeion Handwritten Digit dataset [85]. In addition, CMFL can achieve slightly higher accuracy since it involves the elimination of irrelevant updates that are outliers that harm training.

As a summary, we present the discussed works in this subsection in Table 1.3.

1.4 Resource Allocation

FL involves the participation of heterogeneous workers that have different dataset qualities, computation capabilities, energy states, and willingness to participate. Given the worker heterogeneity and resource constraints, i.e., in device energy states and communication bandwidth, resource allocation has to be optimized to maximize the efficiency of the training process. In particular, the following resource allocation issues need to be considered:

Table 1.3 Approaches to communication cost reduction in FL

Approaches	Ref.	Key ideas	Tradeoffs and shortcomings
Edge and end computation	[21]	More local updates in between communication for global aggregation, to reduce instances of communication	Increased computation cost and poor performance in non-IID setting
	[73]	Similar to the ideas of [21], but with convergence guarantees for vertical FL	Increased computation cost and delayed convergence if global aggregation is too infrequent
	[74]	Transfer learning-inspired two-stream model for FL workers to learn from the fixed global model so as to accelerate training convergence	Increased computation cost and delayed convergence
	[78]	MEC-inspired edge server-assisted FL that aids in intermediate parameter aggregation to reduce instances of communication	System model is not scalable when there are more edge servers
Model compression	[65]	Structured and sketched updates to compress local models communicated from worker to FL server	Model accuracy and convergence issues
	[70]	Similar to the ideas of [65], but for communication from FL server to workers	Model accuracy and convergence issues

(contiuned)

Table 1.3 (continued)

Approaches	Ref.	Key ideas	Tradeoffs and shortcomings
Importance-based updating	[71]	Selective communication of gradients that are assigned importance scores, i.e., to reduce training loss	Only empirically tested on simple datasets and tasks, with fluctuating results
	[67]	Selective communication of local model updates that have higher relevance scores when compared to previous global model	Difficult to implement when global aggregations are less frequent

1. *Worker Selection:* As part of the FL protocol presented in Sect. 1.2.3, worker selection refers to the selection of devices to participate in each training round. Typically, a set of workers is randomly selected by the server to participate. Then, the server has to aggregate parameter updates from all participating devices in the round before taking a weighted average of the models [21]. As such, the training progress of FL is limited by the training time of the slowest participating devices, i.e., stragglers [86]. New worker selection protocols are thus investigated in order to address the training bottleneck in FL.
2. *Joint Radio and Computation Resource Management:* Even though computation capabilities of mobile devices have grown rapidly, many devices still face a scarcity of radio resources [87]. Given that local model transmission is an integral part of FL, there has been a growing number of studies that focus on developing novel wireless communication techniques for efficient FL.
3. *Adaptive Aggregation:* As discussed in Sect. 1.2.1, FL involves global aggregation in which model parameters are communicated to the FL server for aggregation. The conventional approach to global aggregation is a synchronous one, i.e., global aggregations occur in fixed intervals after all workers complete a certain number of rounds of local computation. However, adaptive calibrations of global aggregation frequency can be investigated to increase training efficiency subject to resource constraints [86].
4. *Incentive Mechanism:* In the practical implementation of FL, workers may be reluctant to participate in a federation without receiving compensation since training models are resource consuming. In addition, there exists information asymmetry between the FL server and workers since workers have greater knowledge of their available computation resources and data quality. Therefore, incentive mechanisms have to be carefully designed to both incentivize participation and reduce the potential adverse impacts of information asymmetry.

1.4.1 Worker Selection

To mitigate the training bottleneck, the authors in [88] propose a new FL protocol called *FedCS*. This protocol is illustrated in Fig. 1.4. The system model is an MEC framework in which the operator of the MEC is the FL server that coordinates training in a cellular network that comprises participating mobile devices that have heterogeneous resources. Accordingly, the FL server first conducts a *Resource Request* step to gather information such as wireless channel states and computing capabilities from a subset of randomly selected workers. Based on this information, the MEC operator selects the maximum possible number of workers that can complete the training within a prespecified deadline for the subsequent global aggregation phase. By selecting the maximum possible number of workers in each round, accuracy and efficiency of training are preserved. To solve the maximization problem, a greedy algorithm [89] is proposed, i.e., workers that take the least time

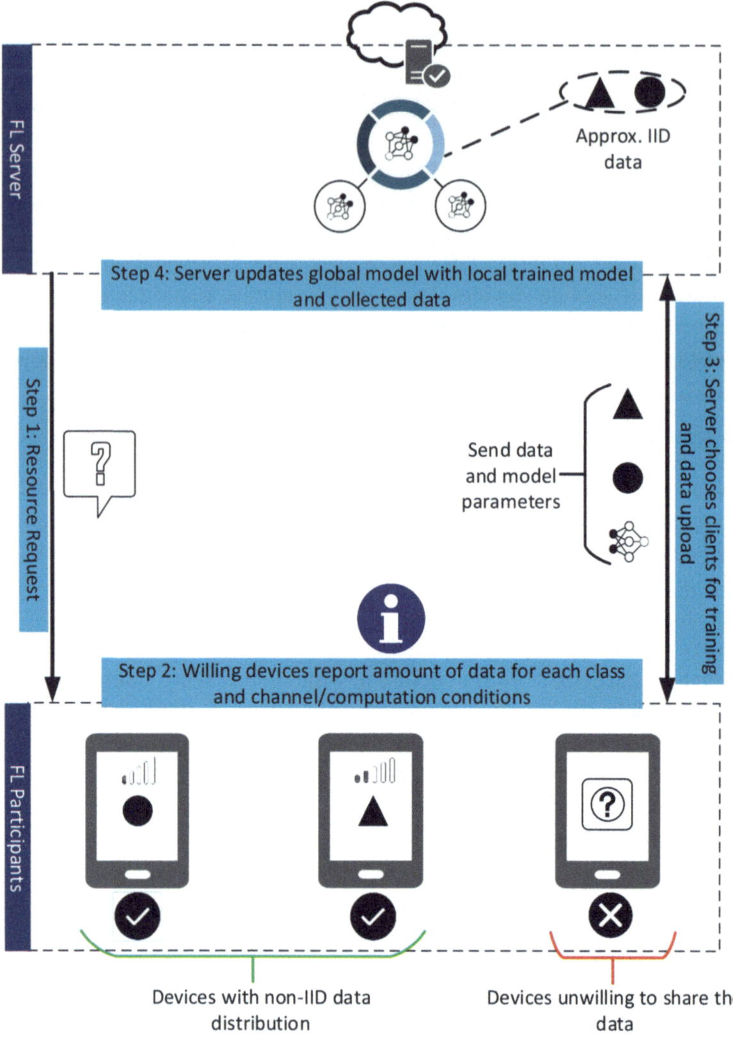

Fig. 1.4 Worker selection under the FedCS and Hybrid-FL protocol

for model upload and update are iteratively selected for training. The simulation results show that compared with the FL protocol which only accounts for training deadline without performing worker selection, *FedCS* can achieve higher accuracy since *FedCS* is able to involve more workers in each training round [21]. However, *FedCS* has been tested only on simple DNN models. When extended to the training of more complex models, it may be difficult to estimate how many workers should be selected. In addition, there is a bias toward selecting workers with devices that have better computing capabilities. These workers may not hold data that is representative of the population distribution. In particular, we revisit the fairness issue [90] subsequently in this section.

While *FedCS* addresses heterogeneity of resources among workers in FL, the authors in [91] extend their work on the *FedCS* protocol with the Hybrid-FL protocol that deals with differences in data distributions among workers. The dataset of workers participating in FL may be non-IID since it is reflective of each individual user's specific characteristics. As we have discussed in Section 1.2.2, the non-IID dataset may significantly degrade the performance of the *FedAvg* algorithm [46]. One proposed measure to address the non-IID nature of the dataset is to distribute publicly available data to workers, such that the EMD between their on-device dataset and the population distance is reduced. However, such a dataset may not always exist, and the workers may not download them for security reasons. Thus, an alternative solution is to construct an approximately IID dataset using inputs from a limited number of privacy insensitive workers [91]. In the Hybrid-FL protocol, during the *Resource Request* step (Fig. 1.4), the MEC operator asks random workers if they permit their data to be uploaded. During the worker selection phase, apart from selecting workers based on computing capabilities, the workers are selected such that their uploaded data can form an approximately IID dataset in the server, i.e., the amount of collected data in each class has close values (Fig. 1.4). Thereafter, the server trains a model on the collected IID dataset and merges this model with the global model trained by workers. The simulation results show that even with just 1% of workers sharing their data, classification accuracy for non-IID data can be significantly improved as compared to the aforementioned *FedCS* benchmark where data are not uploaded at all. However, the recommended protocol can violate the privacy and security of users, especially if the FL server is malicious. In addition, the proposed measure can be costly especially in the case of videos and images. As such, it is unlikely that workers will voluntarily upload data when they can free ride on the efforts of other volunteers. For feasibility, well-designed incentive and reputation mechanisms are needed to ensure that only trustworthy workers are allowed to upload their data.

In general, the mobile edge network environment in which FL is implemented is dynamic and uncertain with variable constraints, e.g., wireless network and energy conditions. Thus, this can lead to training bottlenecks. To this end, Deep Q-Learning (DQL) can be used to optimize resource allocation for model training as proposed in [92]. The system model is a Mobile Crowd Machine Learning (MCML) setting that enables workers in a mobile crowd network to collaboratively train DNN models required by an FL server. The participating mobile devices are constrained by energy, CPU, and wireless bandwidth. Thus, the server needs to determine the proper amounts of data, energy, and CPU resources that the mobile devices use for training to minimize energy consumption and training time. Under the uncertainty of the mobile environment, a stochastic optimization problem is formulated. In the problem, the server is the agent, the state space includes the CPU and energy states of the mobile devices, and the action space includes the number of data units and energy units taken from the mobile devices. To achieve the objective, the reward function is defined as a function of the accumulated data, energy consumption, and training latency. To overcome the large state and action space issues of the server, the DQL technique based on Double Deep Q-Network (DDQN) [93] is adopted to solve the server's problem. The simulation results show that the DQL scheme can reduce energy consumption by around 31% compared with the greedy algorithm, and training latency is reduced up to 55% compared with the random scheme. However, the proposed scheme is applicable only in federations with few participating mobile devices.

The aforementioned resource allocation approaches focus on improving the training efficiency of FL. However, this may cause some FL workers to be left out of the aggregation phase because they are stragglers with limited computing or communication resources.

A consequence of this is unfair resource allocation, a topic that is commonly explored in resource allocation for wireless networks [94] and ML [95]. For example, if the worker selection protocol selects mobile devices with higher computing capabilities to participate in each training round [88], the FL model will be over-represented by the distribution of data owned by workers with devices that have higher computing capabilities. Therefore, the authors in [90] and [96] consider fairness as an additional objective in FL. Fairness is defined in [96] to be the *variance* of performance of an FL model across workers. If the variance of the testing accuracy is large, this implies the presence of more bias or less fairness, since the learned model may be highly accurate for certain workers and less so for other underrepresented workers. The authors in [96] propose the q-Fair FL (q-FFL) algorithm that reweighs the objective function in *FedAvg* to assign higher weights in the loss function to devices with higher loss. The modified objective function is as follows:

$$\min_{w} F_q(w) = \sum_{k=1}^{m} \frac{p_k}{q+1} F_k^{q+1}(w), \tag{1.4}$$

where F_k refers to the standard loss functions presented in Table 1.2, q refers to the calibration of fairness in the system model, i.e., setting $q = 0$ returns the formulation to the typical FL objective, and p_k refers to the ratio of local samples to the total number of training samples. In fact, this is a generalization of the Agnostic FL (AFL) algorithm proposed in [90], in which the device with the highest loss dominates the entire loss function. The simulation results show that the proposed q-FFL can achieve lower variance of testing accuracy and converges more quickly than the AFL algorithm. However, as expected, for some calibrations of the q-FFL algorithm, there can be convergence slowdown since stragglers can delay the training process. As such, an asynchronous aggregation approach can potentially be considered for use with the q-FFL algorithm. As an alternative, the authors in [97] propose a neural network based approach to estimate the local models of FL workers that are left out during training, so as to reduce inference bias.

1.4.2 Joint Radio and Computation Resource Management

While most FL studies have previously assumed orthogonal-access schemes such as orthogonal frequency-division multiple access (OFDMA) [98], the authors in [99] propose a multi-access Broadband Analog Aggregation (BAA) design for the communication latency reduction in FL. Instead of performing communication and computation separately during global aggregation at the server, the BAA scheme builds on the concept of over-the-air computation [100] to *integrate* computation and communication through exploiting the signal superposition property of a multiple access channel. The proposed BAA scheme allows the reuse of the whole bandwidth (Fig. 1.5a), whereas OFDMA orthogonalizes bandwidth allocation (Fig. 1.5b). As such, for orthogonal-access schemes, communication latency increases in direct proportion with the number of workers, whereas for multi-access schemes, latency is independent of the number of workers. The bottleneck of signal-to-noise ratio (SNR) during BAA transmission is the participating device with the longest propagation distance given that devices that are nearer have to lower their transmission power for amplitude alignment with devices located further. To increase SNR, workers with longer propagation distances have to be dropped. However, this leads to the truncation of model parameters. As such, to manage the SNR-truncation tradeoff, three scheduling schemes are considered, namely (a) *cell-interior scheduling*: workers beyond a distance threshold are not scheduled, (b) *all-inclusive scheduling*: all workers are considered, and (c) *alternating scheduling*: edge server alternates between the two aforementioned schemes. The simulation results show that the proposed BAA scheme can achieve similar test accuracy as the OFDMA scheme while achieving latency reduction from 10 times to 1000 times. As a comparison between the three scheduling schemes, the cell-interior scheme outperforms the all-inclusive scheme in terms of test accuracy for high mobility networks where workers have rapidly changing locations. For low mobility networks, the alternating scheduling scheme outperforms cell-interior scheduling.

As an extension, the authors in [101] also introduce error accumulation and gradient sparsification in addition to over-the-air computation. In [99], gradient vectors that are not transmitted as a result of power constraints are completely dropped. To improve the model accuracy, the untransmitted gradient vectors can first be stored in an error accumulation vector. In the next round, local gradient estimates are then corrected using the error vector. In addition, when there are bandwidth limitations, the participating device can apply gradient sparsification to keep only elements with the highest magnitudes for transmission. The elements that are not transmitted are subsequently added to the error accumulation vector for gradient estimate correction in the next round. The simulation results show that the proposed scheme can achieve higher test accuracy than over-the-air computation without error accumulation or gradient sparsification since it corrects gradient estimates with the error accumulation vector and allows for more efficient utilization of the bandwidth.

Similar to [99] and [101], the authors in [102] propose an integration of computation and communication via over-the-air computation. However, it is

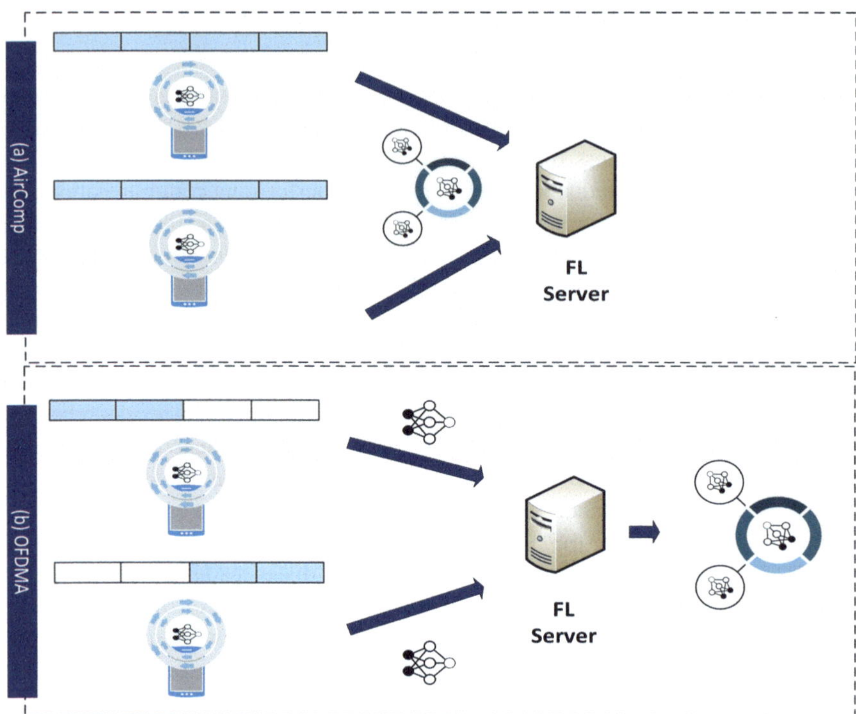

Fig. 1.5 A comparison between (**a**) BAA by over-the-air computation that reuses bandwidth (above) and (**b**) OFDMA (below) that uses only the allocated bandwidth

observed that aggregation error incurred during over-the-air computation can lead to a drop in model accuracy [103] as a result of signal distortion. As such, a worker selection algorithm is proposed in which the number of devices selected for training is maximized so as to improve statistical learning performance [21] while keeping the signal distortion below a threshold. Due to the nonconvexity [104] of the mean square error (MSE) constraint and intractability of the optimization problem, a difference-of-convex functions (DC) algorithm [105] is proposed to solve the maximization problem. The simulation results show that the proposed DC algorithm is scalable and can also achieve near-optimal performance that is comparable to global optimization, which is non-scalable due to its exponential time complexity. In comparison with other state-of-the-art approaches such as the semidefinite relaxation (SDR) technique proposed in [106], the proposed DC algorithm can also select more workers, thus also achieving higher model accuracy.

1.4.3 Adaptive Aggregation

The proposed *FedAvg* algorithm synchronously aggregates parameters as shown in Fig. 1.6a and is thus susceptible to the straggler effect, i.e., each training round only progresses as fast as the slowest device since the FL server waits for *all* devices to complete local training before global aggregation can take place [86]. In addition, the model does not account for workers that can join halfway when the training round is already in progress. As such, the asynchronous model is proposed to improve the scalability and efficiency of FL. For asynchronous FL, the server updates the global model whenever it receives a local update (Fig. 1.6b). The authors in [86] find empirically that an asynchronous approach is robust to workers joining halfway during a training round, as well as when the federation involves participating devices with heterogeneous processing capabilities. However, the model convergence is found to be significantly delayed when data are non-IID

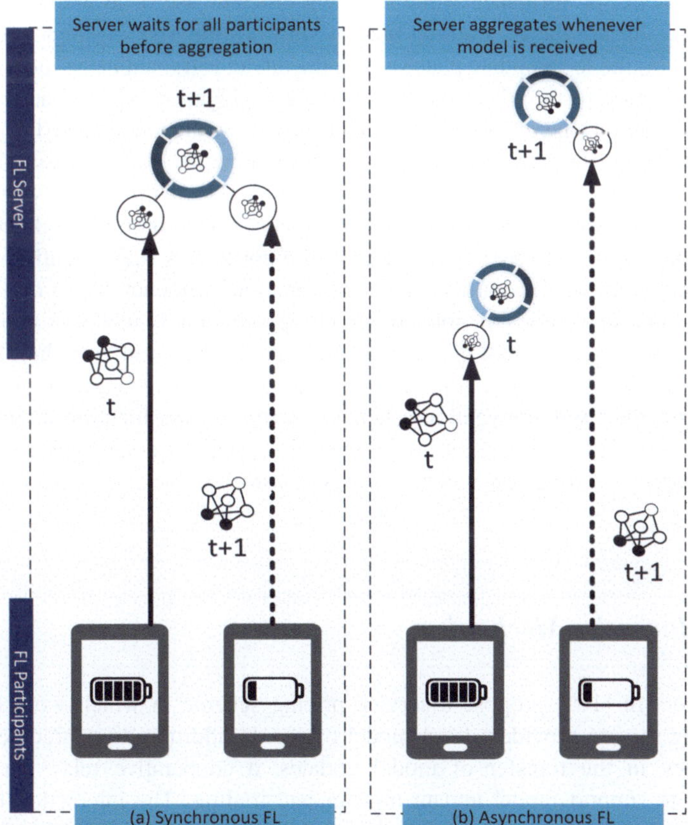

Fig. 1.6 A comparison between (**a**) synchronous and (**b**) asynchronous FL

and unbalanced. As an improvement, [107] proposes the *FedAsync* algorithm in which newly received local updates are adaptively weighted according to staleness, which is defined as the difference between the current epoch and iteration in which the received update belongs to. For example, a stale update from a straggler is outdated since it should have been received in previous training rounds. As such, it is weighted less. In addition, the authors also prove the convergence guarantee for a restricted family of non-convex problems. However, the current hyperparameters of the *FedAsync* algorithm still have to be tuned to ensure convergence in different settings. As such, the algorithm is still unable to generalize to suit the dynamic computation constraints of heterogeneous devices. In fact, given the uncertainty surrounding the reliability of asynchronous FL, synchronous FL remains to be the approach most commonly used today [108].

For most existing implementations of the *FedAvg* algorithm, the global aggregation phase occurs after a fixed number of training rounds. To better manage the dynamic resource constraints, the authors in [44] propose an adaptive global aggregation scheme that varies the global aggregation frequency so as to ensure desirable model performance while ensuring efficient use of available resources, e.g., energy, during the FL training process. In [44], the MEC system model used consists of (a) the local update phase where the model is trained using local data, (b) edge aggregation phase where the intermediate aggregation occurs, and (c) global aggregation phase where the updated model parameters are received and aggregated by the FL server. In particular, the authors study how the training loss is affected when the total number of edge server aggregations and local updates between global aggregation intervals varies. For this, a convergence bound of gradient descent with non-IID data is first derived. Then, a control algorithm is subsequently proposed to adaptively choose the optimal global aggregation frequency based on the most recent system state. For example, if global aggregation is too time consuming, more edge aggregations will take place before communication with the FL server is initiated. The simulation results show that the adaptive aggregation scheme outperforms the fixed aggregation scheme in terms of loss function minimization and accuracy within the same time budget. However, the convergence guarantee of the adaptive aggregation scheme is only considered for convex loss functions currently.

1.4.4 Incentive Mechanism

The authors in [109] propose a service pricing scheme in which workers serve as training service providers for a model owner. In addition, to overcome energy inefficiency in the transfer of model updates, a cooperative relay network is proposed to support model update transfer and trading. The interaction between workers and model owner is modeled as a Stackelberg game [110] in which the model owner is the buyer and workers are the sellers. In the Stackelberg game, each rational worker can noncooperatively decide on its own profit maximization price. In the lower-level subgame, the model owner determines the size of training data to maximize profits considering the increasing concave relationship between the learning accuracy of the model and the size of training data. In the upper-level subgame, the workers decide the price per unit of data to maximize their individual profits. The simulation results show that the proposed mechanism can ensure the uniqueness of the Stackelberg equilibrium. For example, model updates that contain valuable information are priced higher at the Stackelberg equilibrium. In addition, model updates can be transferred cooperatively, thus reducing congestion in communication and improving energy efficiency. However, the simulation environment only involves relatively few mobile devices.

Similar to [109], the authors in [111, 112] also model the interaction between workers and model owner as a Stackelberg game, which is well suited to represent the FL server–worker interaction involved in FL.

However, unlike the aforementioned conventional approaches to solving Stackelberg formulations, a DRL based approach is adopted together with the Stackelberg game by the authors in [113]. In the DRL formulation, the FL server acts as an agent that decides a payment in response to the participation level and payment history of edge nodes, with the objective of minimizing incentive expenses. Then, the edge nodes determine an optimal participation level in response to the payment policy. This learning based incentive mechanism design enables the FL server to derive an optimal policy in response to its observed state, without requiring any prior information.

In contrast to [109, 111–113], the authors in [114] propose an incentive design using a contract theoretic [115] approaches to attract workers with high-quality data for FL. In particular, well-designed contracts can reduce information asymmetry through self-revealing mechanisms in which workers select only the contracts specifically designed for their types. For feasibility, each contract must satisfy the Individual Rationality (IR) and Incentive Compatibility (IC) constraints. For IR, each worker is assured of a positive utility when the worker participates in the federation. For IC, every utility maximizing worker only chooses the contract designed for its own type. The model owner aims to maximize its own profits subject to IR and IC constraints. As illustrated in Fig. 1.7, the optimal contracts derived are self-revealing such that each high-type worker with higher data quality only chooses contracts designed for its type, whereas each low-type worker with lower data quality does not have the incentive to imitate high-type workers. The simulation results show that all types of workers only achieve maximum utility when they choose the contract that matches their types. In addition, the proposed contract theory approach also has better performance in terms of profit for the model owner compared with the Stackelberg game-based incentive mechanism. This is because under the contract theoretic approach, the model owner can extract more profits from the workers, whereas under the Stackelberg game approach, the workers can optimize their individual utilities. In fact, the information asymmetry between FL servers and workers makes contract theory a powerful and efficient tool for mechanism design in FL.

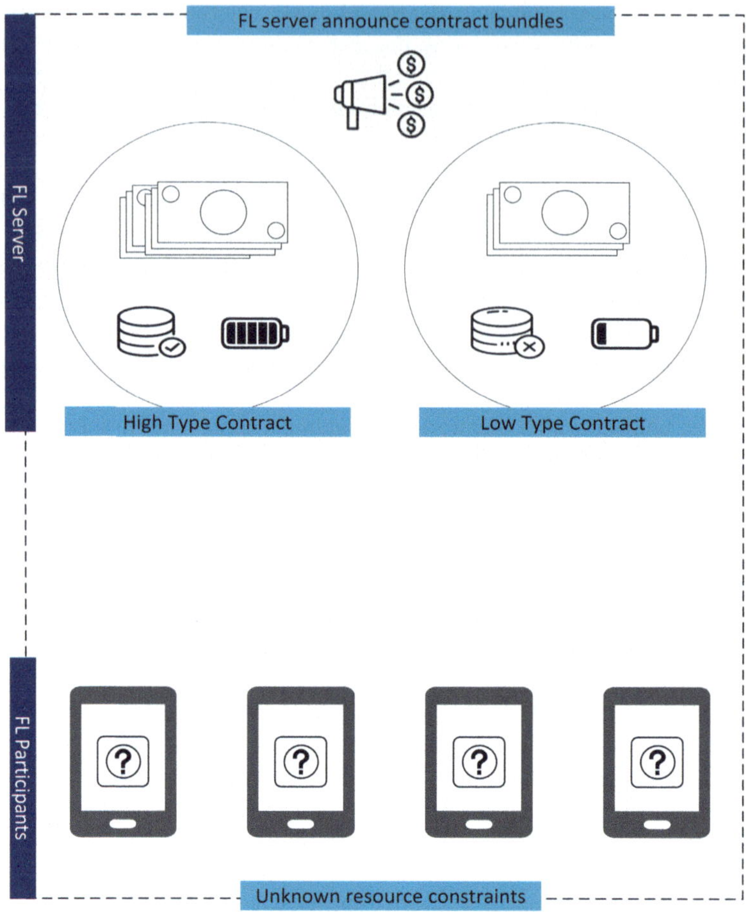

Fig. 1.7 Workers with unknown resource constraints maximize their utility only if they choose the bundle that best reflects their constraints

The authors in [114] further introduce reputation as a metric to measure the reliability of FL workers and design a reputation-based worker selection scheme for reliable FL [41]. In this setting, each worker has a reputation value [116] derived from two sources: (a) direct reputation opinions from past interactions with the FL server and (b) indirect reputation opinions from other task publishers, i.e., other FL servers. The indirect reputation opinions are stored in an open-access reputation blockchain [117] to ensure secure reputation management in a decentralized manner. Before model training, the workers choose a contract that best fits its dataset accuracy and resource conditions. Then, the FL server chooses the workers that have reputation scores that are larger than a prespecified threshold. After the FL task is completed, i.e., a desirable accuracy is achieved, the FL server updates the reputation opinions, which are subsequently stored in the reputation blockchain. The simulation results show that the proposed scheme can significantly improve the accuracy of the FL model since unreliable workers are detected and not selected for FL training.

As a summary, we present the works reviewed in this chapter in Table 1.4.

Table 1.4 Approaches to resource allocation in FL

Approaches	Ref.	Key ideas	Tradeoffs and shortcomings
Worker selection	[88]	FedCS to select workers based on computation capabilities so as to complete FL training before specified deadline	Difficult to estimate training duration accurately for complex models
	[91]	Following [88], Hybrid-FL to select workers so as to accumulate IID, distributable data for FL model training	Request of data sharing may defeat the original intent of FL
	[92]	DRL to determine resource consumption by FL workers	DRL models are difficult to train especially when the number of FL workers are large
	[120]	Following [92], DRL for resource allocation with mobility-aware FL workers	
	[96]	Fair resource allocation to reduce variance of model performance	Convergence delays with more fairness
Joint radio and computation resource management	[99]	Propose BAA to integrate computation and communication through exploiting the signal superposition property of multiple-access channel	Signal distortion can lead to drop in accuracy, the scalability is also an issue when large heterogeneous networks are involved
	[101]	Improves on [99] by accounting for gradient vectors that are not transmitted due to power constraints	
	[102]	Improves on [99] using the DC algorithm to minimize aggregation error	
Adaptive aggregation	[86]	Asynchronous FL where model aggregation occurs whenever local updates are received by FL server	Significant delay in convergence in non-IID and unbalanced dataset

(continued)

Table 1.4 (coninued)

Approaches	Ref.	Key ideas	Tradeoffs and shortcomings
	[44]	Adaptive global aggregation frequency based on resource constraints	Convergence guarantees are limited to restrictive assumptions
Incentive mechanism	[109, 111–113]	Stackelberg game for incentivizing higher quantities of training data or compute resource contributed	FL server derives lower profits. Also, assumption that there is only one FL server
	[114]	Contract theoretic approach to incentivize FL workers	Assumption that there is only one FL server
	[41]	Reputation mechanism to select effective workers	

1.5 Privacy and Security Issues

One of the main objectives of FL is to protect the privacy of workers, i.e., the workers only need to share parameters of the trained model instead of sharing their actual data. However, some recent research works have shown that privacy and security concerns may arise when the FL workers or FL servers are malicious in nature. In particular, this defeats the purpose of FL since the resulting global model can be corrupted, or the workers may even have their privacy compromised during model training. In this section, we discuss the following issues:

1. *Privacy*: Even though FL does not require the exchange of data for collaborative model training, a malicious worker can still infer sensitive information, e.g., gender, occupation, and location, from other workers based on their shared models. For example, in [118], when training a binary gender classifier on the FaceScrub [119] dataset, the authors show that they can infer if a certain worker's inputs are included in the dataset just from inspecting the shared model, with a very high accuracy of up to 90%. Thus, in this section, we discuss privacy issues related to the shared models in FL and review solutions proposed to preserve the privacy of workers.
2. *Security*: In FL, the workers locally train the model and share trained parameters with other workers in order to improve the accuracy of prediction. However, this process is susceptible to a variety of attacks, e.g., data and model poisoning, in which a malicious worker can send incorrect parameters or corrupted models to falsify the learning process during global aggregation. Consequently, the global model will be updated incorrectly, and the whole learning system becomes corrupted. This section discusses more details on emerging attacks in FL as well as some recent countermeasures to deal with such attacks.

1.5.1 Privacy Issues

1.5.1.1 Information Exploiting Attacks in Machine Learning: A Brief Overview

One of the first research works that shows the possibility of extracting information from a trained model is [121]. In this paper, the authors show that during the training phase, the correlations implied in the training samples are gathered inside the trained model. Thus, if the trained model is released, it can lead to an unexpected information leakage to attackers. For example, an adversary can infer the ethnicity or gender of a user from its trained voice recognition system. In [122], the authors develop a model inversion algorithm which is very effective in exploiting information from decision tree-based or face recognition trained models. The idea of this approach is to compare the target feature vector with each of the possible values and then derive a weighted probability estimation which is the correct value. The experiment results reveal that by using this technique, the adversary can reconstruct an image of the victim's face from its label with a very high accuracy.

Recently, the authors in [123] show that it is even possible for an adversary to infer information of a victim through queries to the prediction model. In particular, this occurs when a malicious worker has the access to make prediction queries on a trained model. Then, the malicious worker can use the prediction queries to extract the trained model from the data owner. More importantly, the authors point out that this kind of attack can successfully extract model information from a wide range of training models such as decision trees, logistic regressions, SVMs, and even complex training models including DNNs. Some recent research works have also demonstrated the vulnerabilities of DNN-based training models against model extraction attacks [124–126]. Therefore, this raises a serious privacy concern for workers in sharing training models in FL.

1.5.1.2 Differential Privacy-Based Protection Solutions for FL Workers

In order to protect the privacy of parameters trained by DNNs, the authors in [20] introduce a technique, called *differentially private stochastic gradient descent*, which can be effectively implemented on DL algorithms. The key idea of this technique is to add some "noise" to the trained parameters by using a differential privacy-preserving randomized mechanism [127], e.g., a Gaussian mechanism, before sending such parameters to the server. In particular, at the gradient averaging step of a normal FL worker, a Gaussian distribution is used to approximate the differentially private stochastic gradient descent. Then, during the training phase, the worker keeps calculating the probability that malicious workers can exploit information from its shared parameters. Once a predefined threshold is reached, the worker will stop its training process. In this way, the worker can mitigate the risk of revealing private information from its shared parameters.

Inspired by this idea, the authors in [128] develop an approach which can achieve a better privacy protection solution for workers. In this approach, the authors propose two main steps to process data before sending trained parameters to the server. In particular, for each learning round, the aggregate server first selects a random number of workers to train the global model. Then, if a worker is selected to train the global model in a learning round, the worker will adopt the method proposed in [20], i.e., using a Gaussian distribution to add noise to the trained model before sending the trained parameters to the server. In this way, a malicious worker cannot infer information of other workers by using the parameters of shared global model as it has no information regarding who has participated in the training process in each learning round.

1.5.1.3 Collaborative Training Solutions

While DP solutions can protect private information of a honest worker from other malicious workers in FL, they only work well if the server is trustful. If the server is malicious, it can result in a more serious privacy threat to all workers in the network. Thus, the authors in [129] introduce a collaborative DL framework to render multiple workers to learn the global model without uploading their explicit training models to the server. The key idea of this technique is that instead of uploading the whole set of trained parameters to the server and updating the whole global parameters to its local model, each worker wisely selects the number of gradients to upload and the number of parameters from the global model to update as illustrated in Fig. 1.8. In this way, malicious workers cannot infer explicit information from the shared model. One interesting result of this paper is that even when the workers do not share all trained parameters and do not update all parameters from the shared model, the accuracy of proposed solution is still close to that of the case when the server has all dataset to train the global model. For example, for the MNIST dataset [1], the accuracy of prediction model when the workers agree to share 10% and 1% of their parameters are, respectively, 99.14% and 98.71%, compared with 99.17% for the centralized solution when the server has full data to train. However, the approach is yet to be tested on more complex classification tasks.

Although selective parameter sharing and DP solutions can make information exploiting attacks more challenging, the authors in [130] show that these solutions are susceptible to a new type of attack, called powerful attack, developed based on Generative Adversarial Networks (GANs) [131]. GANs is a class of ML technique which uses two neural networks, namely generator network and discriminator network, that compete with each other to train data. The generator network tries to generate the fake data by adding some "noise" to the real data. Then, the generated fake data are passed to the discriminator network for classification. After the training process, the GANs can generate new data with the same statistics as the training dataset. Inspired by this idea, the authors in [130] develop a powerful attack which allows a malicious worker to infer sensitive information from a victim worker even with just a part of shared parameters from the victim as illustrated in Fig. 1.9. To deal

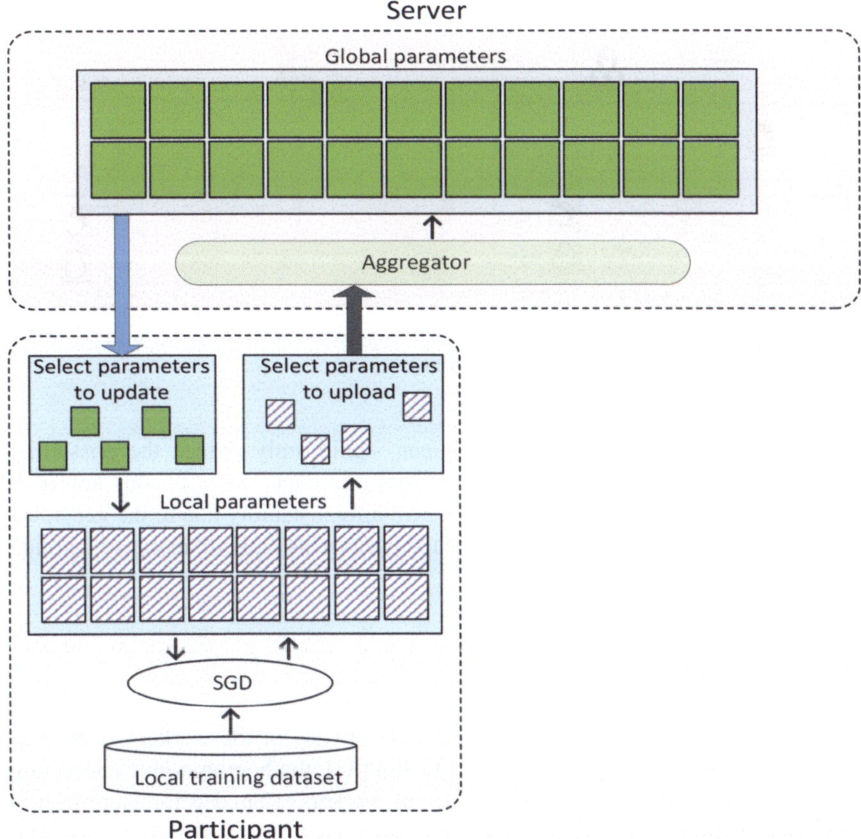

Fig. 1.8 Selective parameter sharing model

with the GAN attack, the authors in [132] introduce a solution using secret sharing scheme with extreme boosting algorithm. This approach executes a lightweight secret sharing protocol before transmitting the newly trained model in plaintext to the server at each round. Thereby, other workers in the network cannot infer information from the shared model. However, the limitation of this approach is the reliance on a trusted third party to generate signature key pairs.

Different from all aforementioned works, the authors in [133] introduce a collaborative training model in which all workers cooperate to train a federated GAN model. The key idea of this method is that the federated GAN model can generate artificial data that can replace workers' real data, thus protecting the privacy of real data for the honest workers. In particular, in order to guarantee workers' data privacy while still maintaining flexibility in training tasks, this approach produces a federated generative model. This model can output artificial data that does not belong to any real user in particular but comes from the common cross-user data

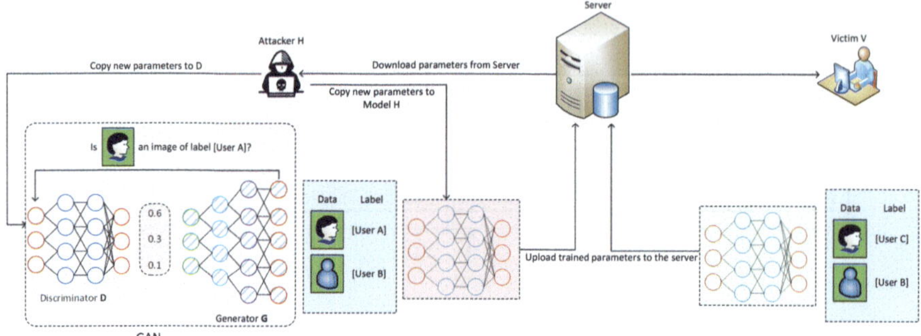

Fig. 1.9 GAN attack on collaborative deep learning

distribution. As a result, this approach can significantly reduce the possibility of malicious exploitation of information from real data. However, this approach inherits existing limitations of GANs, e.g., training instability due to the generated fake data, which can dramatically reduce the performance of collaborative learning models.

1.5.1.4 Encryption-Based Solutions

Encryption is an effective way to protect data privacy of the workers when they want to share the trained parameters in FL. In [134], the homomorphic encryption technique is introduced to protect privacy of workers' shared parameters from a honest-but-curious server. A honest-but-curious server is defined to be a user who wants to extract information from the workers' shared parameters but keeps all operations in FL in a proper working condition. The idea of this solution is that the workers' trained parameters will be encrypted using the homomorphic encryption technique before they are sent to the server. This approach is effective in protecting sensitive information from the curious server and also achieves the same accuracy as that of the centralized DL algorithm. A similar concept is also presented in [135] with secret sharing mechanism used to protect information of FL workers.

Although both the encryption techniques presented in [134] and [135] can prevent the curious server from extracting information, they require multi-round communications and cannot preclude collusions between the server and workers. Thus, the authors in [136] propose a hybrid solution which integrates both additively homomorphic encryption and DP in FL. In particular, before the trained parameters are sent to the server, they will be encrypted using the additively homomorphic encryption mechanism together with intentional noises to perturb the original parameters. As a result, this hybrid scheme can simultaneously prevent the curious server from exploiting information as well as solve the collusion problem between the server and malicious workers. However, in this paper, the authors do not compare the accuracy of the proposed approach with the case without homomorphic encryption and DP. Thus, the performance of proposed approach, i.e., in terms of model accuracy, is not clear.

1.5.2 Security Issues

1.5.2.1 Data Poisoning Attacks

In FL, a worker trains its data and sends the trained model to the server for further processing. In this case, it is intractable for the server to check the real training data of a worker. Thus, a malicious worker can poison the global model by creating *dirty-label* data to train the global model with the aim of generating falsified parameters. For example, a malicious worker can generate a number of samples, e.g., photos, under a designed label, e.g., a clothing branch, and use them to train the global model to achieve its business goal, e.g., the prediction model shows results of the targeted clothing branch. Dirty-label data poisoning attacks are demonstrated to achieve high misclassifications in DL processes, up to 90%, when a malicious worker injects relatively few dirty-label samples (around 50) to the training dataset [137]. This calls for urgent solutions to deal with data poisoning attacks in FL.

In [138], the authors investigate impacts of a sybil-based data poisoning attack to an FL system. In particular, for the sybil attack, a malicious worker tries to improve the effectiveness of data poisoning in training the global model by creating multiple malicious workers. In Table 1.5, the authors show that with only two malicious workers, the attack success rate can achieve up to 96.2%, and now the FL model is unable to correctly classify the image of “1” (instead it always incorrectly predicts them to be the image of “7”). To mitigate sybil attacks, the authors then propose a defense strategy, namely FoolsGold. The key idea of this approach is that honest workers can be distinguished from sybil workers based on their updated gradients. Specifically, in the non-IID FL setting, each worker’s training data has its own particularities, and sybil workers will contribute gradients that appear more similar to each other than those of other honest workers. With FoolsGold, the system can defend the sybil data poisoning attack with minimal changes to the conventional FL process and without requiring any auxiliary information outside of the learning process. Through simulation results on 3 diverse datasets (MNIST [1], KDDCup [139], and Amazon Reviews [139]), the authors show that FoolsGold can mitigate the attack under a variety of conditions, including different distributions of worker data, varying poisoning targets, and various attack strategies.

Table 1.5 The accuracy and attack success rates for no-attack scenario and attacks with 1 and 2 sybils in an FL system with MNIST dataset [1]

	Baseline	Attack 1	Attack 2
Number of honest workers	10	10	10
Number of sybil workers	0	1	2
The accuracy (digits: 0, 2–9)	90.2%	89.4%	88.8%
The accuracy (digit: 1)	96.5%	60.7%	0.0%
Attack success rate	0.0%	35.9%	96.2%

1.5.2.2 Model Poisoning Attacks

Unlike data poisoning attacks that aim to generate fake data to cause adverse impacts to the global model, a model poisoning attack attempts to directly poison the global model that it sends to the server for aggregation. As shown in [140] and [141], model poisoning attacks are much more effective than those of data poisoning attacks, especially for large-scale FL with many workers. The reason is that for data poisoning attacks, a malicious worker's updates are scaled based on its dataset and the number of workers in the federation. However, for model poisoning attacks, a malicious worker can modify the updated model, which is sent to the server for aggregation, directly. As a result, even with one single attacker, the whole global model can be poisoned. The simulation results in [140] also confirm that even a highly constrained adversary with limited training data can achieve high success rate in performing model poisoning attacks. Thus, solutions to protect the global model from model poisoning attacks have to be developed.

In [140], some solutions are suggested to prevent model poisoning attacks. Firstly, based on an updated model shared from a worker, the server can check whether the shared model can help to improve the global model's performance or not. If not, the worker will be marked to be a potential attacker, and after few rounds of observing the updated model from this worker, the server can determine whether this is a malicious worker or not. The second solution is based on the comparison among the updated models shared by the workers. In particular, if an updated model from a worker is too different from the others, the worker can potentially be a malicious one. Then, the server will continue observing updates from this worker before it can determine whether this is a malicious user or not. However, model poisoning attacks are extremely difficult to prevent because when training with millions of workers, it is intractable to evaluate the improvement from every single worker. As such, more effective solutions need to be further investigated.

In [141], the authors introduce a more effective model poisoning attack which is demonstrated to achieve 100% accuracy on the attacker's task within just a single learning round. In particular, a malicious worker can share its poisoned model which not only is trained for its intentional purpose but also contains a backdoor function. In this paper, the authors consider to use a semantic backdoor function to inject into the global model. The reason is that this function can make the global model misclassify even without a need to modify the input data of the malicious worker. For example, an image classification backdoor function can inject an attacker-chosen label to all images with some certain features, e.g., all dogs with black stripes can be misclassified to be cats. In the simulations, the authors show that this attack can greatly outperform other conventional FL data poisoning attacks. For example, in a word-prediction task with 80,000 total workers, compromising just eight of them is enough to achieve 50% backdoor accuracy, as compared to 400 malicious workers needed to perform the data poisoning attack.

1.5.2.3 Free-Riding Attacks

Free-riding is another attack in FL that occurs when a worker wants to benefit from the global model without contributing to the learning process. The malicious worker, i.e., free-rider, can pretend that it has a very small number of samples to train or it can select a small set of its real dataset to train, e.g., to save its resources. As a result, the honest workers need to contribute more resources in the FL training process. To address this problem, the authors in [142] introduce a blockchain-based FL architecture, called BlockFL, in which the workers' local learning model updates are exchanged and verified by leveraging the blockchain technology. In particular, each worker trains and sends the trained global model to its associated miner in the blockchain network and then receives a reward that is proportional to the number of trained data samples as illustrated in Fig. 1.10. In this way, this framework can not only prevent the workers from free-riding but also incentivize all workers to contribute to the learning process. A similar blockchain-based model is also introduced in [143] to provide data confidentiality, computation auditability, and incentives for the workers of FL. However, the utilization of the blockchain technology implies the incurrence of a significant cost for implementing and maintaining miners to operate the blockchain network. Furthermore, consensus protocols used in blockchain networks, e.g., proof of work (PoW), can cause a long delay in information exchange, and thus they may not be appropriate to implement on FL models (Table 1.6).

1.6 Applications of Federated Learning for Mobile Edge Computing

As highlighted by the authors in [33], the increasing complexity and heterogeneity of wireless networks enhance the appeal of adopting a data-driven ML based approach [27] for optimizing system designs and resource allocation decision making for mobile edge networks. However, as discussed in previous sections, the private data of users may be sensitive in nature. As such, existing learning based approach can be combined with FL for privacy-preserving applications.

For brevity, the related studies are not fully reviewed. Table 1.7 provides a brief description of the studies that address the four applications of FL for resource optimization at the wireless edge networks:

1. *Cyberattack Detection:* The ubiquity of IoT devices and increasing sophistication of cyberattacks [144] imply that there is a need to improve existing cyberattack detection tools. Recently, Deep Learning has been widely successful in cyber-attack detection. Coupled with FL, cyberattack detection models can be learned collaboratively while maintaining user privacy.
2. *Edge Caching and Computation Offloading:* Given the computation and storage capacity constraints of edge servers, some computationally intensive tasks of

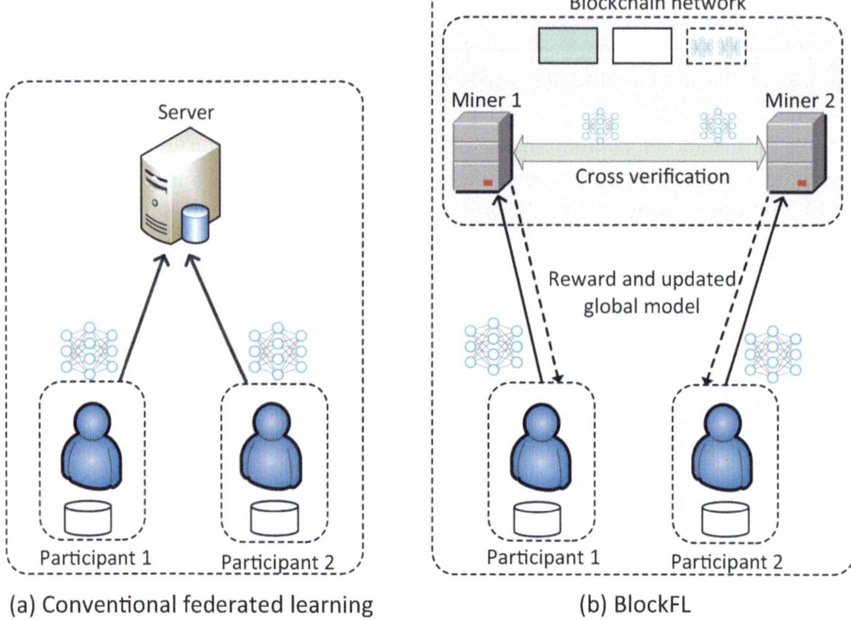

Fig. 1.10 An illustration of (**a**) conventional FL and (**b**) the proposed BlockFL architectures

end devices have to be offloaded to the remote cloud server for computation. In addition, commonly requested files or services should be placed on edge servers for faster retrieval, i.e., users do not have to communicate with the remote cloud when they want to access these files or services. As such, an optimal caching and computation offloading scheme can be collaboratively learned and optimized with FL.

3. *Base Station Association:* In a dense network, it is important to optimize base station association so as to limit interference faced by users. However, traditional learning based approaches that utilize user data often assume that such data are centrally available. Given user privacy constraints, an FL based approach can be adopted instead.
4. *Vehicular Networks:* The Internet of Vehicles (IoV) [145] features smart vehicles with data collection, computation, and communication capabilities for relevant functions, e.g., navigation and traffic management. However, this wealth of knowledge is again private and sensitive in nature since it can reveal the driver's location and personal information. An FL based approach is therefore useful, e.g., in traffic queue length prediction and energy demand in electric vehicle charging stations done at the edge of IoV networks.

Table 1.6 A summary of attacks and countermeasures in FL

Attack types	Attack method	Countermeasures
Information exploiting attacks (privacy issues)	Attackers try to illegally exploit information from the shared model.	• *Differentially private stochastic gradient descent*: Add "noise" to the trained parameters by using a differential privacy-preserving randomized mechanism [20]. • *Differentially private and selective participants*: Add "noise" to the trained parameters and select randomly participants to train global model in each round [128]. • *Selective parameter sharing*: Each participant wisely selects the number of gradients to upload and the number of parameters from the global model to update [129]. • *Secrete sharing scheme with extreme boosting algorithm*: This approach executes a lightweight secret sharing protocol before transmitting the newly trained model in plaintext to the server at each round [132]. • *GAN model training*: All participants are cooperative to train a federated GAN model [133].
Data poisoning attacks	Attackers poison the global model by creating *dirty-label* data and use such data to train the global model.	• *FoolsGoal*: Distinguish honest participants based on their updated gradients. It is based on the fact that in the non-IID FL setting, each participant's training data has its own particularities, and malicious participants will contribute gradients that appear more similar to each other than those of the honest participants [138].
Model poisoning attacks	Attackers attempt to directly poison the global model that they send to the server for aggregation.	• Based on an updated model shared from a participant, the server can check whether the shared model can help to improve the global model's performance or not. If not, the participant will be marked to be a potential attacker [140]. • Compare among the updated global models shared by the participants, and if an updated global model from a participant is too different from others, it could be a potential malicious participant [140].

(continued)

Table 1.6 (continued)

Attack types	Attack method	Countermeasures
Free-riding attacks	Attackers benefit from the global model without contributing to the learning process, e.g., by pretending that they have very small number of samples to train.	• *BlockFL*: Participants' local learning model updates are exchanged and verified by leveraging blockchain technology. In particular, each participant trains and sends the trained global model to its associated miner in the blockchain network and then receives a reward that is proportional to the number of trained data samples [142].

Table 1.7 FL based approaches for mobile edge network optimization

Applications	Ref.	Description
Cyberattack detection	[149]	Cyberattack detection with edge nodes as workers
	[150]	Cyberattack detection with IoT gateways as workers
	[151]	Blockchain to store model updates
Edge caching and computation offloading	[2]	DRL for caching and offloading in UEs
	[152]	DRL for computation offloading in IoT devices
	[153]	Stacked autoencoder learning for proactive caching
	[154]	Greedy algorithm to optimize service placement schemes
Base station association	[155]	Deep echo state networks for VR application
	[30]	Mean-field game with imitation for cell association
Vehicular networks	[31]	Extreme value theory for large queue length prediction
	[156]	Energy demand learning in electric vehicular networks

1.6.1 Cyberattack Detection

Cyberattack detection is one of the most important steps to promptly prevent and mitigate serious consequences of attacks in mobile edge networks. Among different approaches to detect cyberattacks, DL is considered to be the most effective tool to detect a wide range of attacks with high accuracy. In [146], the authors show that DL can outperform all conventional ML techniques with very high accuracy in detecting intrusions on three datasets, i.e., KDDcup 1999, NSL-KDD [147], and UNSW-NB15 [148]. However, the detection accuracy of solutions based on DL

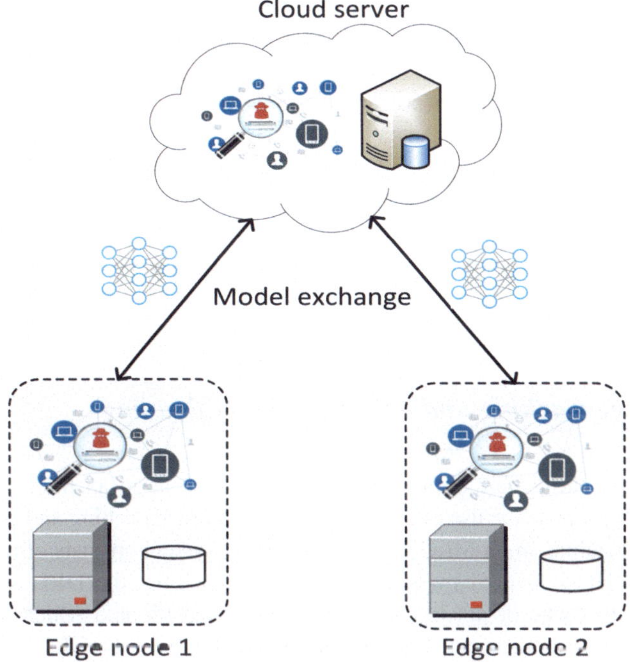

Fig. 1.11 FL based attack detection architecture for IoT edge networks

depends very much on the available datasets. Specifically, DL algorithm only can outperform other ML techniques when given sufficient data to train. However, this data may be sensitive in nature. Therefore, some FL based attack detection models for mobile edge networks have been introduced recently to address this problem.

In [149], the authors propose a cyberattack detection model for an edge network empowered by FL. In this model, each edge node operates as a worker who owns a set of data for intrusion detection. To improve the accuracy in detecting attacks, after training the global model, each worker will send its trained model to the FL server. The server will aggregate all parameters from the workers and send the updated global model back to all the workers as illustrated in Fig. 1.11. In this way, each edge node can learn from other edge nodes without a need of sharing its real data. As a result, this method can not only improve accuracy in detecting attacks but also enhance the privacy of intrusion data at the edge nodes and reduce traffic load for the whole network. A similar idea is also presented in [150] in which IoT gateways operate as FL workers and an IoT security service provider works as a server node to aggregate trained models shared by the workers. The authors in [150] show empirically that by using FL, the system can successfully detect 95.6% of attacks in approximately 257 ms without raising any false alarm when evaluated in a real-world smart home deployment setting.

In both [149] and [150], it is assumed that the workers, i.e., edge nodes and IoT gateways, are honest, and they are willing to contribute to training their updated model parameters. However, if some of the workers are malicious, they can make the whole intrusion detection corrupted. Thus, the authors in [151] propose to use blockchain technology in managing data shared by the workers. By using the blockchain, all incremental updates to the anomaly detection ML model are stored in the ledger, and thus a malicious worker can be easily identified. Furthermore, based on shared models from honest workers stored in the ledger, the intrusion detection system can easily recover the proper global model if the current global model is poisoned.

1.6.2 Edge Caching and Computation Offloading

To account for the dynamic and time-varying conditions in an MEC system, the authors in [2] propose the use of DRL with FL to optimize caching and computation offloading decisions in an MEC system. The MEC system consists of a set of user equipments (UEs) covered by base stations. For caching, the DRL agent makes the decision to cache or not to cache the downloaded file and which local file to replace should caching occur. For computation offloading, the UEs can choose to either offload computation tasks to the edge node via wireless channels or perform the tasks locally. This caching and offloading decision process is illustrated in Fig. 1.12. The states of the MEC system include wireless network conditions, UE energy consumption, and task queuing states, whereas the reward function is defined as quality of experience (QoE) of the UEs. Given the large state and action space in the MEC environment, a DDQN approach is adopted. To protect the privacy of users, an FL approach is proposed in which training can occur with data remaining on the UEs. In addition, existing FL algorithms, e.g., *FedAvg* [21], can also ensure that training is robust to the unbalanced and non-IID data of the UEs. The simulation results show that the DDQN with FL approach achieves similar average utilities among UEs as compared to the centralized DDQN approach, while consuming less communication resources and preserving user privacy. However, the simulations are only performed with 10 UEs. If the implementation is expanded to target a larger number of heterogeneous UEs, there can be significant delays in the training process especially since the training of a DRL model is computationally intensive. As an extension, transfer learning [157] can be used to increase the efficiency of training, i.e., training is not initialized from scratch.

Similar to [2], the authors in [152] propose the use of DRL in optimizing computation offloading decisions in IoT systems. The system model consists of IoT devices and edge nodes. The IoT devices can harvest energy units [158] from the edge nodes to be stored in the energy queue. In addition, an IoT device also maintains a local task queue with unprocessed and unsuccessfully processed tasks. These tasks can be processed locally or offloaded to the edge nodes for processing, in a First In First Out (FIFO) order [15]. In the DRL problem formulation, the

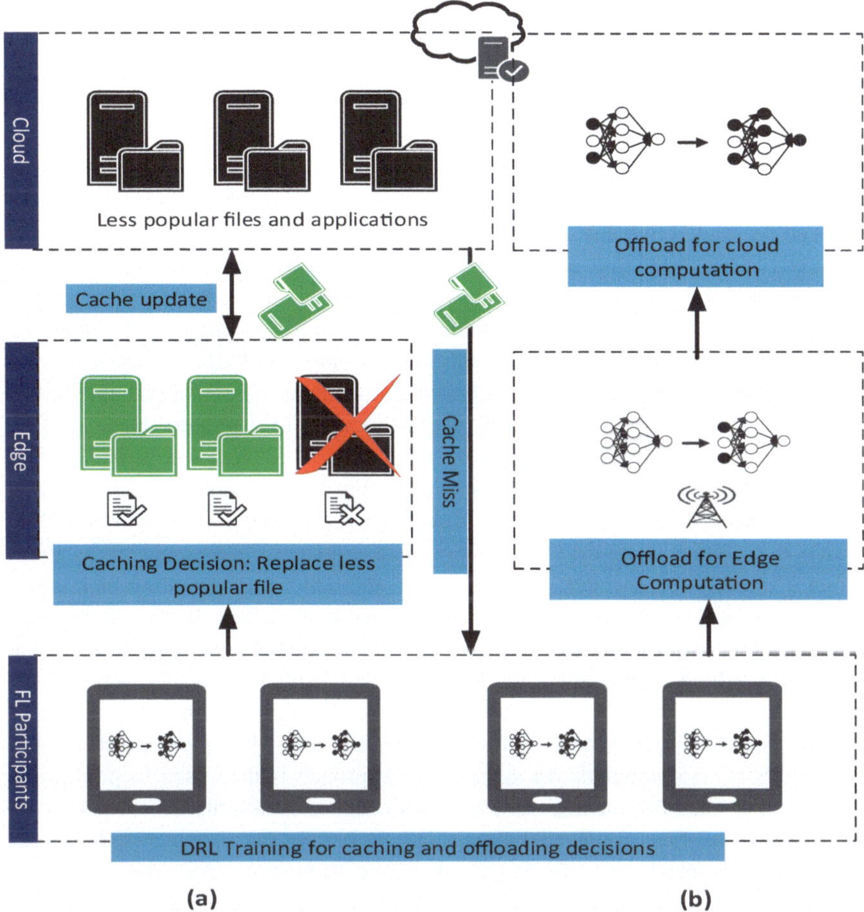

Fig. 1.12 FL based (**a**) caching and (**b**) computation offloading

network states are defined to be a function of energy queue length, task execution delay, task handover delay from edge node association, and channel gain between the IoT device and edge nodes. A task can fail to be executed, e.g., when there is insufficient energy units or communication bandwidth for computation offloading. The utility considered is a function of task execution delay, task queuing delay, number of failed tasks, and penalty of execution failure. The DRL agent makes the decision to either offload computation to the edge nodes or perform computation locally. To ensure privacy of users, the agent is trained without users having to upload their own data to a centralized server. In each training round, a random set of IoT devices is selected to download the model parameters of the DRL agent from the edge networks. The model parameters are then updated using their own data, e.g., energy resource level, channel gain, and local sensing data. Then, the updated parameters of the DRL agent are sent to the edge nodes for model aggregation. The simulation results show that the FL based approach can achieve same levels of total utility as the centralized DRL approach. This is robust to varying task generation probabilities. In addition, when task generation probabilities are higher, i.e., there are more tasks for computation in the IoT device, the FL based scheme can achieve a lower number of dropped tasks and shorter queuing delay than the centralized DRL scheme. However, the simulation only involves 15 IoT devices serviced by relatively many edge nodes. To better reflect practical scenarios where fewer edge nodes have to cover several IoT devices, further studies can be conducted on optimizing the edge-IoT collaboration. For example, the limited communication bandwidth can cause significant task handover delay during computation offloading. In addition, with more IoT devices, the DRL training will take a longer time to converge especially since the devices have heterogeneous computation capabilities.

Instead of using a DRL approach, the authors in [153] propose the use of an FL based stacked autoencoder learning model, i.e., FL based proactive content caching (FPCC) scheme, to predict content popularity for optimized caching while protecting user privacy. In the system model, each user is equipped with a mobile device that connects to the base station that covers its geographical location. Using a stacked autoencoder learning model, the latent representation of a user's information, e.g., location, and file rating, i.e., content request history, is learned. Then, a similarity matrix between the user and its historically requested files is obtained in which each element of the matrix represents the distance between the user and the file. Based on this similarity matrix, the K nearest neighbors of each user are determined, and the similarity between the user's historical watch list and the neighbors' is computed. An aggregation approach is then used to predict the most popular files for caching, i.e., files with highest similarity scores across all users. Being the most popular files across users that are most frequently retrieved, the cached files need not be re-downloaded from its source server every time it is demanded. To protect the privacy of users, FL is adopted to learn the parameters of the stacked autoencoder without the user having to reveal its personal information or its content request history to the FL server. In each training round, the user first downloads a global model from the FL server. Then, the model is trained and updated using their local data. The updated models are subsequently uploaded to the FL server and aggregated using the *FedAvg* algorithm. The simulation results show that the proposed FPCC scheme could achieve the highest cache efficiency, i.e., the ratio of cached files matching user requests, as compared to other caching methods such as the Thompson sampling methods [159]. In addition, privacy of the user is preserved.

The authors in [154] introduce a privacy-aware service placement scheme to deploy user-preferred services on edge servers with consideration for resource constraints in the edge cloud. The system model consists of a mobile edge cloud serving various mobile devices. The user's preference model is first built based on information such as number of times of requests for a service and other user context information, e.g., ages and locations. However, since this can involve sensitive personal information, an FL based approach is proposed to train the preference model while keeping users' data on their personal devices. Then, an optimization problem is formulated in which the objective is to maximize quantity of services demanded from the edge based on user preferences, subject to constraints of storage capacity, computation capability, and bandwidth. The optimization problem is then solved using a greedy algorithm, i.e., the service that most improves the objective function is added till resource constraints are met. The simulation results show that the proposed scheme can outperform the popular service placement scheme, i.e., where only the most popular services are placed on the edge cloud, in terms of the number of requests processed on edge clouds since it also considers the aforementioned resource constraints in maximizing quantity of services.

1.6.3 Base Station Association

The authors in [155] propose an FL based deep echo state networks (ESNs) approach to minimize breaks in presence (BIPs) [160] for users of virtual reality (VR) applications. A BIP event can be a result of delay in information transmission which can be caused when the user's body movements obstruct the wireless link. BIPs cause the user to be aware that they are in a virtual environment, thus reducing their quality of experience. As such, a user association policy has to be designed such that BIPs are minimized. The system model consists of base stations that cover a set of VR users. The base stations receive uploaded tracking information from each associated user, e.g., physical location and orientation, while the users download VR videos for their use in the VR application. For data transmission, the VR users have to associate with one of the base stations. As such, a minimization problem is formulated where BIPs are minimized with respect to expected locations and orientations of the VR user. To derive a prediction of user locations and orientations, the base station has to rely on the historical information of users. However, the historical information stored at each base station only collects partial data from each user, i.e., a user connects to multiple base stations and its data are distributed across them. As such, an FL based approach is implemented whereby each base station first trains a local model using its partial data. Then, the local models are aggregated to form a global model capable of generalization, i.e., comprehensively predicting a user's mobility and orientations. The simulation results show that the federated ESN algorithm can achieve lower BIPs experienced by users as compared to the centralized ESN algorithm proposed in [161], since a centralized approach only makes partial prediction with the incomplete data from sole base stations, whereas the federated ESN approach can make predictions based on a model learned collaboratively from more complete data.

Following the ubiquity of IoT devices, the traditional cloud-based approach may no longer be sufficient to cater to dense cellular networks. As computation and storage operations are moved closer to the edge of the network the association of users with base stations is increasingly important to facilitate efficient ML model training among the end users. To this end, the authors in [30] consider solving the problem of cell association in dense wireless networks with a collaborative learning approach. In the system model, the base stations cover a set of users in an LTE cellular system. In a cellular system, the users are likely to face similar channel conditions as their neighbors and thus can benefit from learning from their neighbors that are already associated with base stations. As such, the cell association problem is formulated as a mean-field game (MFG) with imitation [162] in which each user maximizes its own throughput while minimizing the cost of imitation. The MFG is further reduced into a single-user Markov decision process that is then solved by a neural Q-learning algorithm. In most other proposed solution for cell association, it is assumed that all information is known to the base stations and users. However, given privacy concerns, the assumption of information sharing may not be practical. As such, a collaborative learning approach can be considered where only the outcome of the learning algorithm is exchanged during the learning process, whereas usage data is kept locally in each user's own device. The simulation results show that imitating users can attain higher utility within a shorter training duration as compared to non-imitating users.

1.6.4 Vehicular Networks

Ultra-reliable low latency communication (URLLC) in vehicular networks is an essential prerequisite toward developing an intelligent transport system. However, existing radio resource management techniques do not account for rare events such as large queue lengths at the tail-end distribution. To model the occurrence of such low probability events, the authors in [163] propose the use of extreme value theory (EVT) [164]. The approach requires sufficient samples of queue state information (QSI) and data exchange among vehicles. As such, an FL approach is proposed in which vehicular users (VUEs) train the learning model with data kept locally and upload only their updated model parameters to the roadside units (RSUs). The RSU then averages out the model parameters and returns an updated global model to the VUEs. In a synchronous approach, all VUEs upload their models at the end of a prespecified interval. However, the simultaneous uploading by multiple vehicles can lead to delays in communication. In contrast, for an asynchronous approach, each VUE only evaluates and uploads their model parameters after a predefined number of QSI samples are collected. The global model is also updated whenever a local update is received, thus reducing communication delays. To further reduce overhead, Lyapunov optimization [165] for power allocation is also utilized. The simulation results show that under this framework, there is a reduction of the number of vehicles experiencing large queue lengths, whereas FL can ensure minimal data exchange relative to a centralized approach.

Apart from QSI, the vehicles in vehicular networks are also exposed to a wealth of useful captured images that can be adopted to build better inference models, e.g., for traffic optimization. However, these images are sensitive in nature since they can give away the location information of vehicular clients. As such, an FL approach can be used to facilitate collaborative ML while ensuring privacy preservation. However, the images captured by vehicles are often varying in quality due to motion blurs. In addition, another source of heterogeneity is the difference in computing capabilities of vehicles. Given the information asymmetry involved, the authors in [166] propose a multi-dimensional contract design in which the FL server designs contract bundles comprising varying levels of data quality, compute resources, and contractual payoffs. Then, the vehicular client chooses the contract bundle that maximizes its utility, in accordance with its hidden type. Similar to the results in [114], the simulation results show that the FL server derives greatest utility under the proposed contract theoretic approach, in contrast to the linear pricing or Stackelberg game approach.

The authors in [156] propose a federated energy demand learning (FEDL) approach to manage energy resources in charging stations (CSs) for electric vehicles (EVs). When a large number of EVs congregate at a CS, this can lead to energy transfer congestion. To resolve this, energy is supplied from the power grids and reserved in advance to meet the real-time demands from the EVs [167], rather than having the CS request for energy from the power grid only upon receiving charging requests. As such, there is a need to forecast energy demand for EV networks using historical charging data. However, these data are usually stored separately at each of the CS that the EVs utilize and is private in nature. As such, an FEDL approach is adopted in which each CS trains the demand prediction model on its own dataset before sending only the gradient information to the charging station provider (CSP). Then, the gradient information from the CS is aggregated for global model training. To further improve model accuracy, the CSs are clustered using the constrained K-means algorithm [168] based on their physical locations. The clustering-based FEDL reduces the cost of biased prediction [169]. The simulation results show that the root mean squared error of a clustered FEDL model is lower than conventional ML algorithms, e.g., multi-layer perceptron regressor [170]. However, the privacy of user data is still not protected by this approach, since the user data are stored in each of the CS. As an extension, the user data can possibly be stored in each EV separately, and model training can be conducted in the EVs rather than the CSs. This can allow more user features to be considered to enhance the accuracy of EDL, e.g., user consumption habits.

Summary In this section, we discuss that FL can also be used for mobile edge network optimization. In particular, DL and DRL approaches are suitable for modeling the dynamic environment of increasingly complex edge networks but require sufficient data for training. With FL, model training can be carried out while preserving the privacy of users. A summary of the approaches is presented in Table 1.7.

1.7 Conclusion and Chapter Discussion

This chapter has presented a tutorial of FL and a comprehensive review on the issues regarding FL implementation. Firstly, we begin with an introduction to the motivation for MEC and how FL can serve as an enabling technology for collaborative model training at mobile edge networks. Then, we describe the fundamentals of FL model training. Afterward, we provide detailed reviews, analyses, and comparisons of approaches for emerging implementation challenges in FL. Furthermore, we also briefly discuss the implementation of FL for privacy-preserving mobile edge network optimization. Below, we list the lessons learned from our review of the literature, which paves the subsequent chapters ahead in this book.

1. Incentivizing workers is a challenging task. In a large network, the workers have varying costs, which they are not obliged to report truthfully. With the use of self-revealing mechanisms in contract theory [41, 114], the hidden types of workers are accounted in devising incentive schemes for FL. However, existing studies consider only the single dimensional contract, which is insufficient to capture the complexities and heterogeneities required for workers in the network. In Chapter 2, we devise the multi-dimensional contract incentive scheme to account for multiple sources of information asymmetry in an FL network. Specifically, we utilize this mechanism to incentivize FL in the UAV network, which is characterized by heterogeneity in traversal cost, communication cost, and computation costs.
2. Existing studies of incentive mechanism design generally assume that a federation enjoys a monopoly. In particular, each FL network is assumed to only consist of multiple individual workers collaborating with a sole FL server. There can be exceptions to this setting as follows: (a) the workers may be competing data owners who are reluctant to share their model parameters since the competitors also benefit from a trained global model and (b) the FL servers may compete with other FL servers, i.e., model owners. The FL servers may also restrict the participation of workers in more than one FL training. In this case, the formulation of the incentive mechanism design will be vastly different from that proposed one. In Chap. 4, we devise a deep learning based auction mechanism to model the competition among multiple model owners for the restricted participation of model owners.
3. In heterogeneous mobile networks, the consideration of resource allocation is important to ensure efficient FL. In Sect. 1.4, we have explored different dimensions of resource heterogeneity for consideration, e.g., varying computation and communication capabilities, willingness to participate, and quality of data for local model training. In addition, we have explored various tools that can be considered for resource allocation. Naturally, traditional optimization approaches have also been well explored in radio resource management for FL, given the high dependency on communications efficiency in FL. However, the assumption of static system constraints has been common. In fact, most studies do not consider the dynamic and self-organizing aspects of workers in the FL network.

In Chap. 4, we consider the resource optimization in FL networks subjected to *dynamic* system states.

4. In Sect. 1.3, we learn that communication cost reduction using model compression schemes comes with sacrifices in terms of either higher computation costs or lower inference accuracy. Similarly, there exist different tradeoffs to be considered in resource allocation. A scalable model is thus one that enables customization to suit varying needs. For example, the study of [96] allows the FL server to calibrate levels of fairness when allocating training importance, whereas the study in [171] enables the tradeoffs between training completion time and energy expense to be calibrated by the FL system administrator. Apart from working to directly reduce the size of the model communicated, studies on FL can draw inspiration from applications and approaches in the MEC paradigm. In Chap. 3, we leverage UAVs to facilitate the intermediate aggregation phase of FL. In Chap. 4, we propose a hierarchical FL framework that leverages the computation and communication capacities of edge servers toward scalable and efficient FL. This circumvents the need for model compression that results in accuracy loss.

Chapter 2
Multi-dimensional Contract Matching Design for Federated Learning in UAV Networks

2.1 Introduction

Following the advancements in the Internet of Things (IoT) and edge computing paradigm, traditional Vehicular Ad Hoc Networks (VANETs) that focus mainly on Vehicle-to-Vehicle (V2V) and Vehicle-to-Infrastructure (V2I) communications [172, 173] are gradually evolving into the Internet of Vehicles (IoV) paradigm [174, 175].

The IoV is an open and integrated network system which leverages the enhanced sensing, communication, and computation capabilities of its component data sources, e.g., vehicular sensors, IoT devices, and Roadside Units (RSUs) [176], to build data-driven applications for Intelligent Transport Systems, e.g., for traffic prediction [177], traffic management [178], route planning [179], and other smart city applications [180]. Coupled with the rise of Deep Learning, the wealth of data and enhanced computation capabilities of IoV components enable effective Artificial Intelligence (AI) based models to be built.

Beyond ground data sources, aerial platforms are increasingly important today given that modern day traffic networks have grown in complexity. In particular, Unmanned Aerial Vehicles (UAVs) are commonly used today to provide data collection and computation offloading support in the IoV paradigm. The UAVs feature the benefits of high mobility, flexible deployment, and cost effectiveness [181] and can also provide more comprehensive coverage as compared to ground users. UAVs can be deployed, e.g., to capture images of car parks for the management and analysis of parking occupancy [182], to capture images of roads and highways for traffic monitoring applications [183–185], and also to aggregate data from stationary vehicles and roadside units that in turn collect data of other passing vehicles periodically [186]. Apart from data collection, the UAVs have also been used to provide computation offloading support for resource-constrained IoV components [187, 188].

W. Y. B. Lim et al., *Federated Learning Over Wireless Edge Networks*, Wireless Networks, https://doi.org/10.1007/978-3-031-07838-5_2

As such, studies proposing the Internet of Drones (IoD) and Drones-as-a-Service (DaaS) [189–191] have gained traction recently. Moreover, the DaaS industry is a rapidly growing one [192] that comprises independent drone owners that provide on-demand data collection and model training for businesses and city planners.

Naturally, to build a better inference model, the independently owned UAV companies can collaborate by sharing their data collected from various sources, e.g., carparks, RSUs, and highways, for collaborative model training. However, in recent years, the regulations governing data privacy, e.g., GDPR, are increasingly stringent. As such, this can potentially prevent the sharing of data across DaaS providers. To this end, we propose the adoption of FL based [21] approach to enable privacy-preserving collaborative ML across a federation of independent DaaS providers.

In our system model (Fig. 2.1), a client, hereinafter model owner, is interested in collecting data from a region for model training, e.g., for traffic prediction. Given the energy constraints of UAVs [193], the region is further divided into smaller subregions. The model owner then announces an FL task, e.g., the capturing of real-time traffic flow over highways or the collection of data from RSUs for model training [185]. Then, only the DaaS providers, hereinafter UAVs, that are able to complete the task within the stipulated time and energy constraints respond to the model owner. Thereafter, the model owner assigns an optimal UAV to each subregion. After the UAV collects the sensing data, model training takes place on each UAV separately, following which only the updated model parameters are transmitted to the model owner for global aggregation.

Our proposed approach has three advantages. Firstly, the resource-constrained IoV components are aided by the UAV deployment for completion of time-sensitive sensing and model training tasks. Secondly, it preserves the privacy of the UAV-collected data by eliminating the need for data sharing across UAVs. Thirdly, it is communication efficient. The reason is that traditional methods of data sharing will require the raw data to be uploaded to an aggregating cloud server. With FL, only the model parameters need to be transmitted by the UAVs.

However, there exists an incentive mismatch between the model owner and the UAVs. On the one hand, the model owners aim to maximize their profits by selecting the optimal UAVs which can complete the stipulated task at the lowest cost, e.g., in terms of sensing, transmission, and computation costs. On the other hand, the UAVs can take advantage of the information asymmetry and misreport their types so as to seek higher compensation. To that end, we leverage the self-revealing properties of contract theory [115] as an incentive mechanism design to appropriately reward the UAVs based on their actual types. In particular, different from existing works, given the complexity of the sensing and collaborative learning task, we consider a multi-dimensional contract to account for the multi-dimensional sources of heterogeneity in terms of UAV sensing, learning, and transmission capabilities.

After deriving optimal contracts to which the UAVs respond, the possibility that multiple UAVs prefer a particular subregion still remains. To that end, we leverage the Gale–Shapley (GS) [194] matching-based algorithm to assign the optimal UAVs to each subregion.

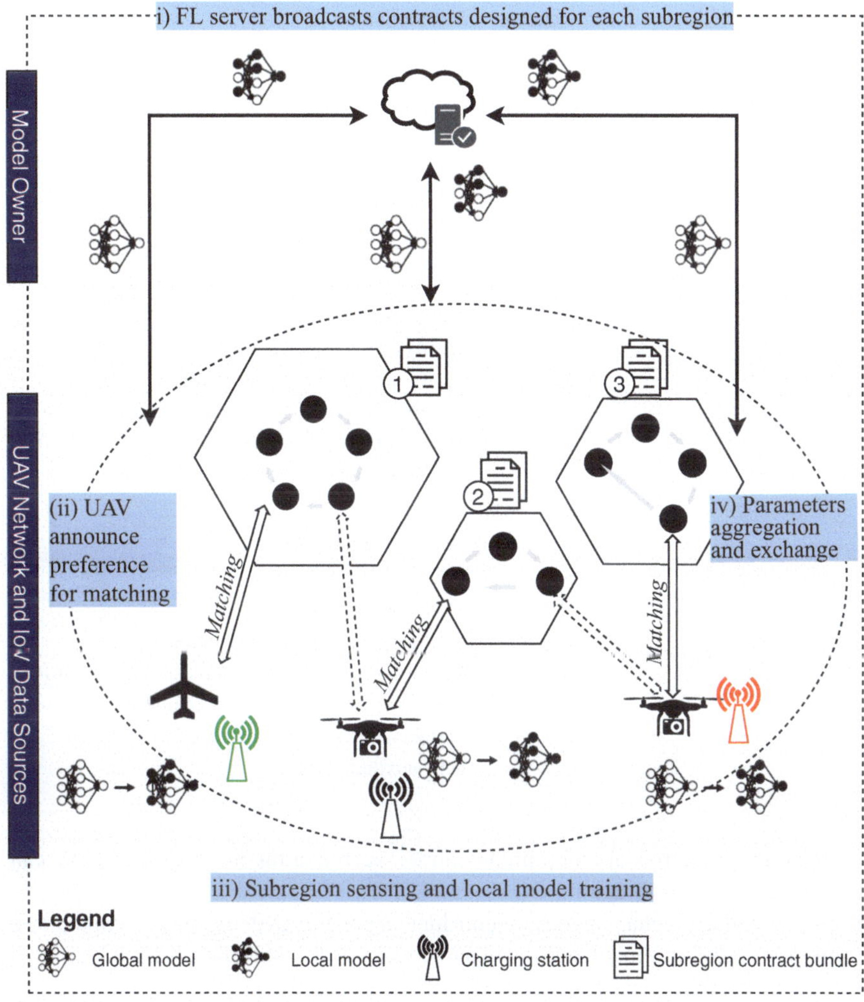

Fig. 2.1 System model involving UAV-subregion contract matching

The contribution of this chapter is as follows:

1. We propose an FL based sensing and collaborative learning scheme in which UAVs collect the data and participate in privacy-preserving collaborative model training for applications in the IoV paradigm toward the development of an Intelligent Transport System.
2. In consideration of the incentive mismatches and information asymmetry between the UAVs and model owner, we propose a multi-dimensional contract matching-based incentive mechanism design that aims to leverage on the self-

Table 2.1 Table of commonly used notations

Notation	Description
n	Subregion
j	UAV
C_j	Base of UAV j
l_j^n	Total sensing distance
$l_{C_j}^n$	Total traversal distance
$\tau_P^{j,n}$	Total duration taken for traversal and sensing
$E_P^{j,n}$	Total energy taken for traversal and sensing
α_j^n	Marginal cost of node coverage for sensing
ψ_j^n	Traversal cost
$\tau_C^{j,n}$	Local computation duration
$E_C^{j,n}$	Total energy taken for computation
β_j	Marginal cost of node coverage for computation
$\tau_T^{j,n}$	Total duration for transmission
ζ_j^n	Energy taken for transmission
u_j^n	UAV utility
R_j^n	Contractual rewards
ϕ	Unit cost of energy for the UAV
Π	Model owner profit
Ω^n, ω^n	Contract set and individual contract
$\tilde{R}$	Compensation for sensing and computation costs
$\hat{R}$	Compensation for traversal and transmission costs
$\upsilon(\alpha_y, \beta_z)$	Marginal cost of node coverage
Φ_i	UAV auxiliary type

revealing properties of an optimal contract, such that the most optimal UAV can be matched to a subregion.

3. Our incentive mechanism design considers a general UAV sensing, computation, and transmission model and thus can be extended to specific FL based applications in the IoV paradigm.

Due to the heavy notations used in this chapter, we provide a list of common notations used in this chapter in Table 2.1. The organization of this chapter is as follows. Section 2.2 introduces the system model and problem formulation, Sect. 2.3 discusses the multi-dimensional contract formulation, Sect. 2.4 considers a matching-based UAV-subregion assignment, Sect. 2.5 presents the performance evaluation of our proposed incentive mechanism design, and Sect. 2.6 concludes.

2.2 System Model and Problem Formulation

We consider a network in which a model owner aims to collect data from stipulated nodes, e.g., from RSUs or images of segments in the highway, in a target sensing region to fulfill a time-sensitive task. One UAV is selected by the task publisher to cover each of the subregions. Given information asymmetry and the multiple sources of heterogeneity in UAV cost types, the model owner leverages the self-revealing properties of a multi-dimensional contract theoretic approach to choose one UAV suited to cover each of the subregion. After data collection, the UAV returns to their respective UAV bases for FL based model training.

Following [195], the target sensing region can be modeled as a graph and divided into N smaller graphs, i.e., subregions whose set is denoted $\mathcal{N} = \{1, \ldots, n, \ldots, N\}$, e.g., through the multilevel graph partition algorithm [196]. The set of nodes in subregion n is denoted $\mathcal{I}_n = \{I_1, \ldots, I_n, \ldots, I_N\}$ with the node i in subregion n located at $\boldsymbol{x}_i^n \in \mathbb{R}^3$. The Euclidean distance between two nodes i and i' located within subregion n, $\forall i, i' \in \mathcal{I}_n, i \neq i'$ is expressed as $l_{i,i'}^n$, where $l_{i,i'}^n = ||\boldsymbol{x}_i^n - \boldsymbol{x}_{i'}^n|| < \infty$, i.e., all nodes are inter-accessible.

A set $\mathcal{J} = \{1, \ldots, j, \ldots, J\}$ of J unmanned aerial vehicles (UAVs) is located at bases situated around the target sensing region. Without loss of generality, we assume that each base owns a single UAV and $J \geq N$. Moreover, our model can be easily extended to scenarios in which a UAV swarm is required for sensing in each subregion. Denote $C_j = \{C_1, \ldots, C_j, \ldots, C_J\}$ as the set of bases, where C_j refers to the base of UAV j located at $\boldsymbol{y}_{C_j} \in \mathbb{R}^3$. The Euclidean distance between the base of UAV j and subregion n is expressed as $l_{C_j}^n$, where $l_{C_j}^n = ||\boldsymbol{y}_{C_j} - \boldsymbol{x}_{\bar{i}}^n|| < \infty$ and $\bar{i}$ denotes the center of the subregion.

There are two stages in our system model as follows:

1. **Multi-dimensional Contract Design:** The UAV types, e.g., sensing, traversal, and transmission costs, are private information not known to the model owner. As such, the model owner designs a multi-dimensional contract to leverage the self-revealing mechanism so as to select the optimal UAV to cover each subregion. In particular, the model owner can maximize its profits by choosing the lowest cost UAV among all feasible UAVs that can complete the task within the time constraint.
2. **UAV-Subregion Assignment:** Each UAV reports its type and ranks the N subregions based on its preferences of coverage to the FL model owner. Then, a stable UAV-subregion matching is derived using the Gale–Shapley (GS) algorithm. Note that each UAV's preference can vary across different subregions. As an illustration, we consider a representative UAV j at base C_j and two subregions n and n' where $l_{C_j}^{n'} \gg l_{C_j}^n, \forall i_n \in I_n$ and $\forall i_{n'} \in I_{n'}$. In this case, UAV j has to traverse a longer distance to reach subregion n'. As such, it is able to cover smaller proportion of subregion n' relative to n due to energy and time constraints.

In the following, we consider the sensing, computation, and data transmission model of a representative UAV.

2.2.1 UAV Sensing Model

We consider a representative UAV j tasked by the model owner to cover a proportion of nodes in the subregion n. Denote the node coverage assignment of UAV j in subregion n to be $\mathcal{A}^{j,n} = \left\{a_{i,i'}^{j,n} | \forall i, i' \in \mathcal{I}_n, i \neq i'\right\}$, where $a_{i,i'}^{j,n} = 1$ represents that the UAV has to fly through the segment between nodes i and i', and $a_{i,i'}^{j,n} = 0$ implies otherwise.

The total distance l_j^n traveled for sensing by UAV j under assignment $\mathcal{A}^{j,n}$ is as follows:

$$l_j^n = \sum_{i' \neq i, i' \in \mathcal{I}_n} a_{i,i'}^{j,n} l_{i,i'}^n. \tag{2.1}$$

Denote $\theta_j^n = \frac{\sum_{i' \neq i, i' \in \mathcal{I}_n} a_{i,i'}^{j,n}}{|\mathcal{I}_n|}$ where $|\cdot|$ indicates cardinality, i.e., θ_j^n refers to the proportion of node coverage by UAV j in subregion n where $0 \leq \theta_j^n \leq 1$.

Apart from traveling between the nodes, the UAV has to travel to and from its base. Denote the total distance traveled by the UAV as $L_j^n = l_j^n + l_{C_j}^n$. Hereinafter, we refer to l_j^n as the *sensing* distance, whereas $l_{C_j}^n$ refers to the *traversal* distance.

Following the works of [197, 198], each UAV travels with an average velocity v_j and expends a fixed propulsion power $p_j = c_{j,1} v_j^3 + \frac{c_{j,2}}{v_j}$ throughout the task for tractability, where $c_{j,1}$ and $c_{j,2}$ refer to the required power to balance the parasitic drag caused by skin friction and required power to balance the drag force of air redirection, respectively.[1] Note that the propulsion power consumed by the UAV when it changes its direction is negligible [193]. The total duration taken for traversal and sensing is denoted $\tau_P^{j,n} = \frac{L_j^n}{v_j}$, whereas the total energy consumed to cover the traversal and sensing distance is as follows:

$$\begin{aligned} E_P^{j,n} &= \frac{L_j^n}{v_j} p_j = \frac{\theta_j^n l^n + l_{C_j}^n}{v_j} p_j \\ &= \frac{p_j l^n}{v_j} \theta_j^n + \frac{l_{C_j}^n}{v_j} p_j \\ &= \alpha_j^n \theta_j^n + \psi_j^n, \end{aligned} \tag{2.2}$$

where l^n is the distance traveled by the UAV if it covers all nodes, i.e., $\theta_j^n = 1$, $\alpha_j^n = \frac{p_j l^n}{v_j}$, and $\psi_j^n = \frac{l_{C_j}^n}{v_j} p_j$ for notation simplicity. Note that α_j^n represents the sensing

[1] In practice, the propulsion power is in turn a function of other factors, e.g., reference area of the UAV and wing aspect ratio and weight. For simplicity, we consider that p_j accounts for these factors.

cost, i.e., marginal cost of node coverage for sensing in the subregion, whereas ψ_j^n refers to the traversal cost, i.e., the energy cost of traveling to and from the base. A higher α_j^n can imply that the UAV j requires greater propulsion power to complete the task, e.g., due to its larger weight or wing aspect ratio, whereas a higher ψ_j^n implies either a greater propulsion power to move or a greater traversal cost, i.e., the subregion is farther away from the base. While the value of α_j^n varies across subregions due to the varying l_n, i.e., the marginal cost of node coverage varies according to the sensing area of the subregion, the *ordering* of the UAV types based on the sensing costs is retained. On the other hand, the order of UAVs by traversal costs varies across subregions, based on the distance between the UAV base and each of the subregions.

2.2.2 UAV Computation Model

After the UAV j covers its assigned set of nodes following assignment $\mathcal{A}^{j,n}$, it returns to the base C_j for an FL based model training over K global iterations where $\mathcal{K} = \{1, \ldots, k, \ldots, K\}$ to minimize the global loss $F^K(\boldsymbol{w})$. Following [65], each training iteration k consists of three steps namely (1) *Local Computation*, i.e., the UAV trains the received global model $\boldsymbol{w}^{(k)}$ locally using the sensing data,[2] (2) *Wireless Transmission*, i.e., the UAV transmits the model parameter update $\boldsymbol{h}_j^{(k)}$ to the model owner, and (iii) *Global Model Parameter Update*, i.e., all parameter updates derived from the N subregions are aggregated to derive an updated global model $\boldsymbol{w}^{(k+1)}$, where $\boldsymbol{w}^{(k+1)} = \cup_{j \in \mathcal{N}}(\boldsymbol{w}_j^{(k)} + \boldsymbol{h}_j^{(k)})$, which is then transmitted back to the UAVs for the $(k+1)$th training iteration.

In general, a series of local model training is performed by the UAV to minimize an L-Lipschitz and γ-strongly convex local loss function G_j up to the target accuracy A^* defined by the model owner to derive the parameter update. Note that a larger value of A^* implies a greater deviation from the optimal value. Moreover, $0 < A^* < 1$, i.e., the local solution $\boldsymbol{h}_j^{(k)}$ does not have to be trained to optimality, e.g., to reduce local computation duration especially for time-sensitive tasks. In particular, following the formulation in [171],

$$\begin{aligned} G_j\left(\boldsymbol{w}^{(k)}, \boldsymbol{h}_j^{(k)}\right) - G_j\left(\boldsymbol{w}^{(k)}, \boldsymbol{h}_j^{(k)*}\right) \\ \leq A^*\left(G_j\left(\boldsymbol{w}^{(k)}, \boldsymbol{0}\right) - G_j\left(\boldsymbol{w}^{(k)}, \boldsymbol{h}_j^{(k)*}\right)\right). \end{aligned} \tag{2.3}$$

[2] In practice, note that the model parameters received may be initialized and trained with a generic dataset offline in the cloud first. This set of model parameters is updated afterward during the FL process.

The FL training is completed after $K = \frac{a}{1-A^*}$ global iterations where $a = \frac{2L^2}{\gamma^2 \xi}$ and $0 \leq \xi \leq \frac{\gamma}{L}$. The total local computation duration $\tau_C^{j,n}$ is as follows:

$$\tau_C^{j,n} = K \left(\frac{V C_j \theta_j^n D^n \log_2(1/A^*)}{f_j} \right), \tag{2.4}$$

whereas the energy consumption of UAV j for computation is as follows:

$$E_C^{j,n} = K \left(\kappa C_j \theta^{j,n} D^n V \log_2(1/A^*) f_j^2 \right) = \beta_j \theta_j^n. \tag{2.5}$$

Note that κ is the effective switched capacitance that depends on the chip architecture [199], C_j is the number of cycles per bit for computing one sample data of UAV j, $\theta^{j,n} D^n$ is the unit of data samples collected by UAV j, $V \log_2(1/A^*)$ refers to the lower bound on the number of local iterations required to achieve local accuracy A^* [171] where $V = \frac{2}{(2-L\delta)\delta\gamma}$, and f_j refers to the computation capacity of the UAV j, measured by CPU cycles per second. For ease of notation, we denote $\beta_j = \kappa K C_j D^n V \log_2(1/A^*) f_j^2$, i.e., a higher β_j implies greater energy cost for computation per additional node coverage. Similar to α_j^n, the value of β_j^n varies across subregions due to the different units of data samples available for computation. However, the ordering of the UAV types based on computation costs is retained.

2.2.3 UAV Transmission Model

After local computation, the wireless transmission takes place from the selected UAVs to the model owner. For simplicity, we denote the achievable rate of the UAV j to be a product of its transmit power ρ_j and a scaling factor λ_j^n that covers other considerations, e.g., bandwidth allocation and channel gain.

The total time $\tau_T^{j,n}$ taken by the UAV to upload parameter update $\boldsymbol{h}_j^{(k)}$ of size H is as follows: $\tau_T^{j,n} = K \frac{H}{\lambda_j^n \rho_j}$. Note that the model upload size is constant regardless of the number of global iterations or quantity of data collected, given the fixed dimensions of the model update. The transmission energy consumption, denoted ζ_j^n, is as follows:

$$E_T^{j,n} = \tau_T^{j,n} \rho_j = \zeta_j^n. \tag{2.6}$$

2.2.4 UAV and Model Owner Utility Modeling

The utility function of a representative UAV j covering subregion n can be expressed as follows:

$$\begin{aligned} u_j^n(\theta_j^n) &= R_j^n(\theta_j^n) - \phi\left(E_P^{j,n} + E_C^{j,n} + E_T^{j,n}\right) \\ &= R_j^n(\theta_j^n) - \phi\left(\alpha_j^n\theta_j^n + \psi_j^n + \beta_j^n\theta_j^n + \zeta_j^n\right), \end{aligned} \tag{2.7}$$

where $R_j^n(\theta_j^n)$ refers to the contractual rewards and ϕ refers to the unit cost of energy.

Following [112, 113, 200], the FL model accuracy $\Upsilon(\sum_{n=1}^{N}\theta_{j*}^n D^n)$ is a concave function of the aggregate data collected across N subregions by the N selected UAVs. In particular, the inference accuracy of the model is improved when more nodes are covered, i.e., a model trained using data across a more comprehensive coverage of classes may be built. Without loss of generality, we consider the aggregate model performance to be an average of node coverage across all regions, analogous to the Federated Averaging algorithm:

$$\Upsilon\left(\sum_{n=1}^{N}\theta_{j*}^n D^n\right) = \frac{1}{N}\sum_{n=1}^{N} log(1+\mu\theta_{j*}^n D^n), \tag{2.8}$$

where $\mu > 0$ is the system parameter. The total profit obtained from all UAVs is thus as follows:

$$\Pi(\Omega) = \sigma\Upsilon\left(\sum_{n=1}^{N}\theta_{j*}^n D^n\right) - \sum_{n=1}^{N} R_{j*}^n, \tag{2.9}$$

where $\sigma > 0$ refers to the conversion parameter from model performance to profits, and the contractual reward expense for each selected UAV is denoted as R_{j*}^n. In the next section, we devise the optimal contract which satisfies the Individual Rationality and Incentive Compatibility constraints.

2.3 Multi-dimensional Contract Design

In this section, we first consider a multi-dimensional contract formulation. To solve the multi-dimensional contract, we sort the UAV types according to an auxiliary variable that reflects the marginal cost of node coverage. Then, we relax the constraints for contract feasibility and include a fixed compensation component for traversal and transmission costs so as to solve for the optimal contract.

2.3.1 Contract Condition Analysis

Given that the sensing cost α, traversal cost ψ, computation cost β, and transmission cost ζ are all private information that are not precisely known by the model owner, we consider the multi-dimensional contract theoretic incentive mechanism design to leverage on its self-revealing properties. The reason is that without the self-revealing properties, the UAV operators may misreport their costs in order to seek overly high compensations.

The UAVs can be classified into different types to characterize their heterogeneity. In particular, the UAVs can be categorized into a set $\Psi = \{\psi_x^n : 1 \leq x \leq X\}$ of X traversal cost types, set $\mathcal{A} = \{\alpha_y^n : 1 \leq y \leq Y\}$ of Y sensing cost types, set $\mathcal{B} = \{\beta_z^n : 1 \leq z \leq Z\}$ of Z computation cost types, and set $\mathcal{C} = \{\zeta_q^n : 1 \leq q \leq Q\}$ of Q transmission cost types.

Without loss of generality, we also assume that the user types are indexed in non-decreasing orders in all four dimensions: $0 < \psi_1 \leq \psi_2 \leq \cdots \leq \psi_X$, $0 < \alpha_1^n \leq \alpha_2^n \leq \cdots \leq \alpha_Y^n$, $0 < \beta_1^n \leq \beta_2^n \leq \cdots \leq \beta_Z^n$, and $0 < \zeta_1^n \leq \zeta_2^n \leq \cdots \leq \zeta_Q^n$. For ease of notation, we represent a UAV of traversal cost type x, sensing cost type y, computation cost type z, and transmission cost type q to be that of type-(x, y, z, q).

To enforce the UAVs to truthfully reveal their private information, we adopt a two-step procedure for the contract design:

1. *Multi-dimensional Contract Design:* We convert the multi-dimensional problem into a single-dimensional contract formulation following the approach in [201]. In particular, we sort the UAVs by an auxiliary, one-dimensional type $\Phi(\alpha_y^n, \beta_z^n)$ in the ascending order based on the marginal cost of node coverage, i.e., sensing and computation cost types. Then, we solve for the optimal contract for each subregion n denoted $\Omega^n(\mathcal{A}, \mathcal{B}) = \{\omega_{y,z}^n : 1 \leq y \leq Y, 1 \leq z \leq Z\}$ where $n \in \mathcal{N}$ to derive the optimal node coverage–contract reward bundle $\{\theta_{y,z}^n, \tilde{R}_{y,z}^n\}$.
2. *Traversal Cost Compensation:* In contrast to existing works on multi-dimensional contracts, the UAVs also incur the additional traversal cost and transmission cost components, both of which are not coupled with the marginal cost of node coverage. In other words, these costs have to be incurred regardless of the number of nodes a UAV decides to cover in the subregion. For each contractual reward, we add in a fixed compensation $\hat{R}$ to derive the final contract bundle $\{\theta_{y,z}^n, (\tilde{R}_{y,z}^n + \hat{R})\}$.

We first discuss the multi-dimensional contract formulation as follows. A contract is feasible only if the Individual Rationality (IR) and Incentive Compatibility (IC) constraints hold simultaneously.

Definition 2.1 Individual Rationality (IR): Each type-(y, z) UAV achieves non-negative utility if it chooses the contract item designed for its type, i.e., contract item $\omega_{y,z}$.

$$u_{y,z}(\omega_{y,z}) \geq 0, 1 \leq y \leq Y, 1 \leq z \leq Z. \tag{2.10}$$

Definition 2.2 Incentive Compatibility (IC): Each type-(y, z) UAV achieves the maximum utility if it chooses the contract item designed for its type, i.e., contract item $\omega_{y,z}$. As such, it has no incentive to choose contracts designed for other types.

$$\begin{aligned} u_{y,z}(\omega_{y,z}) \geq u_{,y,z}(\omega_{y',z'}), 1 \leq y \leq Y, 1 \leq z \leq Z, \\ y \neq y', z \neq z'. \end{aligned} \tag{2.11}$$

The multi-dimensional contract formulation is as follows:

$$\begin{aligned} &\max_{\Omega} \Pi(\Omega^n(\mathcal{A}, \mathcal{B})) \\ &\text{s.t. } (2.10),(2.11). \end{aligned} \tag{2.12}$$

However, the optimization problem in (2.12) involves YZ, i.e., IR constraints and $YZ(YZ-1)$, i.e., IC constraints, all of which are non-convex. Therefore, we first convert the contract into a single-dimensional formulation in the next section.

2.3.2 Conversion into a Single-Dimensional Contract

In order to account for the marginal cost of node coverage, we consider a revised utility $\tilde{u}_{y,z}$ of the UAV type-(y, z) that excludes the traversal and transmission costs as follows:

$$\tilde{u}_{y,z}(\theta(y, z), \tilde{R}_{y,z}) = \eta(\alpha_y, \beta_z) + \tilde{R}_{y,z}, \tag{2.13}$$

where we denote $\eta(\alpha_y, \beta_z) = -\phi(\alpha_y\theta(y, z) + \beta_z\theta(y, z))$ for ease of notation, and $\tilde{R}_{y,z}$ refers to the contractual reward arising from the multi-dimensional contract design. To focus on a representative contract, we drop the n superscripts for now. Given that the ranking of marginal cost types does not change across subregion, note that our contract design is a general one applicable to all subregions.

We derive the marginal cost of node coverage $\upsilon(\alpha_y, \beta_z)$ for the type-(y, z) UAV as follows:

$$\upsilon(\alpha_y, \beta_z) = -\frac{\partial\eta(\alpha_y, \beta_z)}{\partial\theta(y, z)} = \phi(\alpha_y + \beta_z). \tag{2.14}$$

Intuitively, $\frac{\partial \eta(\alpha_y, \beta_z)}{\partial \theta(y,z)} < 0$ since the coverage of an additional node results in the additional expenses of sensing and computation costs. A larger value of $\upsilon(\alpha_y, \beta_z)$ implies a larger marginal cost of node coverage, due to the greater sensing and computation costs incurred for a particular UAV type.

We can now sort the YZ UAVs according to their marginal cost of node coverage in a non-decreasing order as follows:

$$\Phi_1(\theta), \Phi_2(\theta), \ldots, \Phi_i(\theta), \ldots, \Phi_{YZ}(\theta), \tag{2.15}$$

where $\Phi_i(\theta)$ denotes the auxiliary type-$\Phi_i(\theta)$ user. Given the sorting order, the UAV types are in an ascending order based on their marginal cost of node coverage:

$$\upsilon(\theta, \Phi_1) \leq \upsilon(\theta, \Phi_2) \leq \cdots \leq \upsilon(\theta, \Phi_i) \leq \cdots \leq \upsilon(\theta, \Phi_{YZ}). \tag{2.16}$$

Note that for ease of notation, we use type-Φ_i or type-i interchangeably to represent the auxiliary type-i user. In addition, we refer to $\eta_i(\theta_i)$ and $\eta(\theta_i, \Phi_i)$ interchangeably to represent the new ordering subsequently. Similarly, to represent the marginal cost of node coverage, we use $\upsilon_i(\theta_i)$ and $\upsilon(\theta_i, \Phi_i)$. In the next section, we derive the necessary and sufficient conditions for the contract design.

2.3.3 *Conditions for Contract Feasibility*

We derive the necessary conditions to guarantee contract feasibility based on the IR and IC constraints as follows.

Lemma 2.1 *For any feasible contract $\Omega\{\mathcal{A}, \mathcal{B}\}$, we have $\theta_i < \theta_{i'}$ if and only if $\tilde{R}_i < \tilde{R}_i'$, $i \neq i'$.*

Proof We first prove the sufficiency, i.e., if $R_i < R_i' \Rightarrow \theta_i < \theta_{i'}$. From the IC constraint of type-Φ_i UAV, we have

$$\eta(\theta_i, \Phi_i) + \tilde{R}_i \geq \eta(\theta_i, \Phi_{i'}) + \tilde{R}_{i'},$$

$$\eta(\theta_i, \Phi_i) - \eta(\theta_i, \Phi_{i'}) \geq \tilde{R}_{i'} - \tilde{R}_i > 0,$$

which implies

$$\eta(\theta_i, \Phi_i) \geq \eta(\theta_i, \Phi_{i'}).$$

Given that $\frac{\partial \eta(\theta_i, \Phi_i)}{\partial \theta_i} < 0$, we can deduce $\theta_i < \theta_{i'}$.

Next, we prove the necessity, i.e., $\theta_i < \theta_{i'} \Rightarrow \tilde{R}_i < \tilde{R}_i'$. Similarly, we consider the IC constraint of the type-Φ_i UAV:

$$\eta(\theta_i, \Phi_i) + \tilde{R}_i \geq \eta(\theta_{i'}, \Phi_i) + \tilde{R}_{i'},$$

$$\eta(\theta_i, \Phi_i) - \eta(\theta_{i'}, \Phi_i) \geq \tilde{R}_{i'} - \tilde{R}_i.$$

Given $\theta_i < \theta_{i'}$, we deduce $\eta(\theta_i, \Phi_i) < \eta(\theta_{i'}, \Phi_i)$, which follows that $\tilde{R}_{i'} < \tilde{R}_i$. The proof is now completed. □

Lemma 2.2 *Monotonicity: For any feasible contract $\Omega\{\mathcal{A}, \mathcal{B}\}$, if $\upsilon(\theta_i, \Phi_i) > \upsilon(\theta_i, \Phi_{i'})$, it follows that $\theta_i \leq \theta_{i'}$.*

Proof We adopt the proof by contradiction to validate the monotonicity condition. We first assume that there exists $\theta_i > \theta_{i'}$ such that $\upsilon(\theta_i, \Phi_i) > \upsilon(\theta_i, \Phi_{i'})$, i.e., the lemma is incorrect.

We consider the IC constraints for the type Φ_i and $\Phi_{i'}$ UAV:

$$\eta(\theta_i, \Phi_i) + \tilde{R}_i \geq \eta(\theta_{i'}, \Phi_i) + \tilde{R}_{i'}, \tag{2.17}$$

$$\eta(\theta_{i'}, \Phi_{i'}) + \tilde{R}_{i'} \geq \eta(\theta_i, \Phi_{i'}) + \tilde{R}_i. \tag{2.18}$$

Then, we add the constraints together and rearrange the terms to obtain

$$[\eta(\theta_i, \Phi_i) - \eta(\theta_{i'}, \Phi_i)] - [\eta(\theta_i, \Phi_{i'}) - \eta(\theta_{i'}, \Phi_{i'})] \geq 0. \tag{2.19}$$

By the fundamental theorem of calculus, we have

$$\begin{aligned}
&[\eta(\theta_i, \Phi_i) - \eta(\theta_{i'}, \Phi_i)] - [\eta(\theta_i, \Phi_{i'}) - \eta(\theta_{i'}, \Phi_{i'})] \\
&= \int_{\theta_{i'}}^{\theta_i} \frac{\partial \eta(\theta, \Phi_i)}{\partial \theta} d\theta - \int_{\theta_{i'}}^{\theta_i} \frac{\partial \eta(\theta, \Phi_{i'})}{\partial \theta} d\theta \\
&= \int_{\theta_{i'}}^{\theta_i} \left[\frac{\partial \eta(\theta, \Phi_i)}{\partial \theta} - \frac{\partial \eta(\theta, \Phi_{i'})}{\partial \theta} \right] d\theta \\
&= - \int_{\theta_{i'}}^{\theta_i} [\upsilon(\theta, \Phi_i) - \upsilon(\theta, \Phi_{i'})] \, d\theta.
\end{aligned} \tag{2.20}$$

Given (2.12) from Sect. 2.3.1, as well as the assumption $\theta_i > \theta_{i'}$ and $\eta(\theta_i, \Phi_i) > \eta(\theta_i, \Phi_{i'})$, we can deduce that (2.20) is negative, which contradicts with (2.19). As such, there does not exist $\theta_i > \theta_{i'}$ and $\eta(\theta_i, \Phi_i) > \eta(\theta_i, \Phi_{i'})$ for the feasible contract, which confirms that the lemma is correct. The proof is now completed. □

As such, Lemmas 2.1 and 2.2 give us the necessary conditions of the feasible contract in the following theorem.

Theorem 2.1 *A feasible contract must meet the following conditions:*

$$\begin{cases} \theta_1 \geq \theta_2 \geq \cdots \geq \theta_i \geq \cdots \geq \theta_{YZ} \\ \tilde{R}_1 \geq \tilde{R}_2 \geq \cdots \geq \tilde{R}_i \geq \cdots \geq \tilde{R}_{YZ}. \end{cases} \tag{2.21}$$

Next, we further relax the IR and IC constraints. Due to the independence of Φ_i on the contract item $\{\theta, \tilde{R}\}$, i.e., $\Phi_i(\theta, \tilde{R}) = \Phi_i(\theta', \tilde{R}'), \theta \neq \theta', \tilde{R} \neq \tilde{R}'$, the UAV type does not change with the node coverage and contract rewards. In addition, the ordering of the type by marginal costs does not change with the subregion n. As such, we are able to deduce the minimum utility UAV Φ_{max}, i.e., the UAV type that incurs the highest marginal cost of node coverage, as follows:

$$\Phi_{max}(\theta) = \arg\min_{\Phi_i} \tilde{u}\left(\theta, \tilde{R}, \Phi_i\right). \tag{2.22}$$

Intuitively, $\Phi_{max} = \Phi_{YZ}$, i.e., the UAV characterized by $\{\alpha_Y, \beta_Z\}$ is the UAV which incurs the highest marginal cost of node coverage, and hence it is the minimum utility UAV.

Lemma 2.3 *If the IR constraint of the minimum utility UAV type Φ_{YZ} is satisfied, the other IR constraints will also hold.*

Proof From the IC constraint and the sorting order $\eta_1(\theta_1) \geq \cdots \geq \eta_i(\theta_i) \cdots \geq \eta_{YZ}(\theta_{YZ})$, we have the following relation:

$$\eta_i(\theta_i) + \tilde{R}_i \geq \eta_i(\theta_{YZ}) + \tilde{R}_{YZ} \geq \eta_{YZ}(\theta_{YZ}) + \tilde{R}_{YZ} \geq 0. \tag{2.23}$$

As such, as long as the IR constraint of the UAV type Φ_{YZ} is satisfied, it follows that the IR constraints of the other UAVs will also hold. □

Lemma 2.4 *For a feasible contract, if $\omega_{i-1} \overset{PIC}{\Leftrightarrow} \omega_i$ and $\omega_i \overset{PIC}{\Leftrightarrow} \omega_{i+1}$, then $\omega_{i-1} \overset{PIC}{\Leftrightarrow} \omega_{i+1}$.*

Note that the relation $\omega_i \overset{PIC}{\Leftrightarrow} \omega_{i'}$, $i \neq i'$ implies the *Pairwise Incentive Compatibility* (PIC), which is fulfilled under the following condition:

$$\begin{cases} u_i(\omega_i) \geq u_i(\omega_{i'}), \\ u_{i'}(\omega_{i'}) \geq u_{i'}(\omega_i). \end{cases} \tag{2.24}$$

Proof Suppose we have three UAV types Φ_{i-1}, Φ_i, and Φ_{i+1} where $i-1 < i < i+1$. The Local Upward Incentive Constraint (LUIC), i.e., IC constraint between the ith and $(i+1)$th UAV is as follows:

$$\eta(\theta_{i-1}, \Phi_{i-1}) + \tilde{R}_{i-1} \geq \eta(\theta_i, \Phi_{i-1}) + \tilde{R}_i \tag{2.25}$$

$$\eta(\theta_i, \Phi_i) + \tilde{R}_i \geq \eta(\theta_{i+1}, \Phi_i) + \tilde{R}_{i+1}. \tag{2.26}$$

In addition, we consider

$$\begin{aligned}
&[\eta(\theta_{i+1}, \Phi_i) - \eta(\theta_i, \Phi_i)] - [\eta(\theta_{i+1}, \Phi_{i-1}) - \eta(\theta_i, \Phi_{i-1})] \\
&= \int_{\theta_i}^{\theta_{i+1}} \frac{\partial \eta(\theta, \Phi_i)}{\partial \theta} d\theta - \int_{\theta_i}^{\theta_{i+1}} \frac{\partial \eta(\theta, \Phi_{i-1})}{\partial \theta} d\theta \\
&= \int_{\theta_i}^{\theta_{i+1}} \frac{\partial \eta(\theta, \Phi_i)}{\partial \theta} - \frac{\partial \eta(\theta, \Phi_{i-1})}{\partial \theta} d\theta \\
&= -\int_{\theta_i}^{\theta_{i+1}} [\upsilon(\theta, \Phi_i) - \upsilon(\theta, \Phi_{i-1})] d\theta \\
&= \int_{\theta_{i+1}}^{\theta_i} [\upsilon(\theta, \Phi_i) - \upsilon(\theta, \Phi_{i-1})] d\theta.
\end{aligned} \tag{2.27}$$

Given the order of marginal cost of node coverage in (2.16) in Sect. 2.3.2, it follows that the LUIC is positive. As such, we have

$$\eta(\theta_{i+1}, \Phi_i) - \eta(\theta_i, \Phi_i) \geq \eta(\theta_{i+1}, \Phi_{i-1}) - \eta(\theta_i, \Phi_{i-1}). \tag{2.28}$$

Adding the LUIC inequalities presented in (2.26) together with that of (2.28), we have

$$\eta(\theta_{i-1}, \Phi_{i-1}) + \tilde{R}_{i-1} \geq \eta(\theta_{i+1}, \Phi_{i-1}) + \tilde{R}_{i+1}. \tag{2.29}$$

By considering the Local Downward Incentive Constraint (LDIC), i.e., IC constraint between the ith and $(i-1)$th UAV, we are able to derive that

$$\eta(\theta_{i+1}, \Phi_{i+1}) + \tilde{R}_{i+1} \geq \eta(\theta_{i-1}, \Phi_{i+1}) + \tilde{R}_{i-1}. \tag{2.30}$$

As such, given that the LUIC and LDIC hold, we have proven that the PIC of the contracts holds, i.e., $\omega_{i-1} \overset{PIC}{\Leftrightarrow} \omega_{i+1}$. □

With Lemma 2.3, we are able to reduce YZ IR constraints into a single constraint, i.e., as long as the minimum utility type Φ_{YZ} UAV has a non-negative utility, it follows that the other IR constraints will hold. Moreover, with Lemma 2.4, we are able to reduce $YZ(YZ-1)$ constraints into $YZ-1$ constraints, i.e., as long as the PIC constraint of the type Φ_i and type Φ_{i+1} UAV holds, it follows that the IC constraints between the type Φ_i and all other UAV types will hold.

With this, we are able to derive a tractable set of sufficient conditions for the feasible contract in Theorem 2.2 as follows. The first condition refers to the reduced IR condition corresponding to Lemma 2.3, whereas the second condition refers to the PIC condition between the type Φ_i and type Φ_{i+1} UAV corresponding to Lemma 2.4.

Theorem 2.2 A feasible contract must meet the following sufficient conditions:

1. $\eta_{YZ}(\theta_{YZ}) + \tilde{R}_{YZ} \geq 0$
2. $\tilde{R}_{i+1} + \eta(\theta_{i+1}, \Phi_{i+1}) - \eta(\theta_i, \Phi_{i+1}) \geq \tilde{R}_i \geq \tilde{R}_{i+1} + \eta(\theta_{i+1}, \Phi_i) - \eta(\theta_i, \Phi_i)$.

2.3.4 Contract Optimality

To solve for the optimal contract rewards $\tilde{R}_i^*$, we first establish the dependence of optimal contract rewards $\mathbf{R}$ on route coverage θ. Thereafter, we solve the problem in (2.12) from Sect. 2.3.1 with $\boldsymbol{\theta}$ only. Specifically, we obtain the optimal rewards $R^*(\boldsymbol{\theta})$ given a set of feasible node coverages from each UAV which satisfies the monotonicity constraint $\theta_1 \geq \theta_2 \geq \cdots \geq \theta_i \geq \cdots \geq \theta_{YZ}$.

In addition, the multi-dimensional contract formulation that we have thus far only considered the self-revelation for two types, i.e., sensing and computation costs. To account for traversal and transmission cost types, we add an additional fixed compensation $\hat{R}$ into the contract rewards. The traversal cost can be derived from the historical information of the UAV and can be calibrated based on the response that the model owner receives. In the following theorem, we prove that the addition of a fixed reward compensation does not violate the IC constraints, i.e., the self-revealing properties of the contract are still preserved, whereas it is inconsequential even if the IR constraint is violated, given that $\tilde{R}$ has already been designed to sufficiently compensate marginal costs, and only one optimal UAV is required to serve each subregion. The optimal rewarding scheme is summarized as follows.

Theorem 2.3 *For a known set of node coverage $\boldsymbol{\theta}$ satisfying $\theta_1 \geq \theta_2 \geq \cdots \geq \theta_i \geq \cdots \geq \theta_{YZ}$ in a feasible contract, the optimal reward is given by*

$$R_i^* = \begin{cases} \hat{R} - \eta(\theta_i, \Phi_i), & if i = YZ, \\ \hat{R} + \tilde{R}_{i+1} + \eta(\theta_{i+1}, \Phi_i) - \eta(\theta_i, \Phi_i), & otherwise. \end{cases} \tag{2.31}$$

Proof There are two parts for the proof. Firstly, we prove that the reward design for the two-dimensional contract is optimal. Therefore, we adopt the proof by contradiction. We first assume there exists some $\mathbf{R}^\dagger$ that yields greater profit for the model owner, meaning that the theorem is incorrect, i.e., $\Pi(R^\dagger) > \Pi(R^*)$. For simplicity, we need to consider only the reward portion of the model owner's profit function in this proof, i.e., $\sum_{i=1}^{YZ} R_i^\dagger < \sum_{i=1}^{YZ} R_i^*$. This implies there exists at least a $t \in \{1, 2, \ldots, YZ\}$ that satisfies the inequality $R_t^\dagger < R_t^*$.

According to the PIC constraint of Lemma 2.4, we have

$$R_t^\dagger \geq R_{t+1}^\dagger + \eta\left(\theta_{t+1}, \Phi_{t+1}\right) - \eta\left(\theta_t, \Phi_t\right). \tag{2.32}$$

In contrast from Theorem 2.3, we have

$$R_t^* = R_{t+1}^* + \eta\left(\theta_{t+1}, \Phi_{t+1}\right) - \eta\left(\theta_t, \Phi_t\right). \tag{2.33}$$

From (2.32) and (2.33), we can deduce that $R^{\dagger}_{t+1} < R^{*}_{t+1}$. Continuing the process up to $t = YZ$, we have $R^{\dagger}_{YZ} \leq R^{*}_{YZ} = -\eta(\theta_i, \Phi_i)$, which violates the IR constraint. As such, there does not exist the rewards $\mathbf{R}^{\dagger}$ in the feasible contract that yields greater profit for the model owner. Intuitively, the model owner chooses the lowest reward that satisfies the IR and IC constraints for profit maximization.

Secondly, we show that adding a fixed traversal cost reward does not violate the IC constraint. Within a subregion, when we consider the complete utility function of the auxiliary UAV with type i, $i \neq i'$,

$$\begin{aligned} &\eta\left(\theta_i, \Phi_i\right) + \tilde{R}_i + \hat{R} - \psi_i^n - \zeta_i^n \geq \\ &\qquad\qquad \eta\left(\theta_{i'}, \Phi_i\right) + \tilde{R}_i + \hat{R} - \psi_i^n - \zeta_i^n. \end{aligned} \tag{2.34}$$

Intuitively, the traversal and transmission cost is structurally separate from the marginal costs, i.e., sensing and computation cost types of the UAV within a subregion n. As such, the fixed reward terms cancel out and the self-revealing properties of the contract are preserved.

Note that the IR constraint may no longer hold for some i where

$$\psi_i > \eta\left(\theta_i, \Phi_i\right) + \tilde{R}_i + \hat{R}^n. \tag{2.35}$$

However, this is inconsequential given that unlike the conventional contract theoretic formulations, we only require a type of UAV to serve a subregion. Moreover, $\tilde{R}$ is already designed such that the IR constraints hold to compensate marginal costs sufficiently. For certain subregions, $\hat{R}$ can be calibrated upward if no UAV responds to the model owner. □

Following (2.31), we can reexpress the optimal rewards as

$$R_i^* = \hat{R} - \eta(\theta_{YZ}, \Phi_{YZ}) + \sum_{t=i}^{YZ} \Delta_t, \tag{2.36}$$

where $\Delta_{YZ} = 0$, $\Delta_t = \eta(\theta_{i+1}, \Phi_i) - \eta(\theta_i, \Phi_i)$, and $t = 1, 2, \ldots, YZ - 1$.

Unlike conventional contract theoretic formulations, the model owner only requires a single contract per subregion n for the optimal UAV. From (2.16) in Sect. 2.3.2, we can deduce that the optimal type-i^* UAV to serve each subregion is $i^* = \arg\max_{\Phi_i \in \boldsymbol{\Phi}} \Pi(\Omega\{\mathcal{A}, \mathcal{B}\}) = \arg\min_{\Phi_i \in \boldsymbol{\Phi}} \Phi_i = \Phi_1$. Intuitively, for each region, the model owner leverages the self-revealing properties of the multi-dimensional contract formulation to obtain an optimal UAV with the lowest marginal cost of node coverage for profit maximization. In other words, this is the UAV that can cover the largest proportion of the subregion at the lowest cost, among all feasible UAVs that can complete the task. We can substitute the optimal rewards into the profit function of the model owner and rewrite the profit maximization problem as follows:

$$\begin{aligned} &\max_{(R,\theta_{i*}^n)} \Pi(\Omega^n) = \sum_{n=1}^{N} G^n(\theta_1^n), \\ &\text{s.t.} \\ &C1: \theta_1^n \geq \theta_2^n \geq \cdots \geq \theta_i^n \geq \cdots \geq \theta_{YZ}^n, \\ &C2: 0 \leq \theta_1^n \leq 1, \end{aligned} \tag{2.37}$$

where

$$\begin{aligned} G^n &= \frac{\sigma}{N} log(1 + \mu\theta_1^n D^n) - R_1^* \\ &= \frac{\sigma}{N} log(1 + \mu\theta_1^n D^n) - \hat{R} - \phi(\alpha_1^n\theta_1^n + \beta_1\theta_1^n). \end{aligned} \tag{2.38}$$

Note that $C1$ refers to the monotonicity constraint of the contract, whereas $C2$ specifies the upper and lower bounds of the route coverage. Taking the first-order derivative of (2.37) to solve for θ_1^{n*}, we can derive the closed-form solution of the optimal node coverage–contract reward pair as follows:

$$\begin{cases} \theta_1^{n*} = \frac{1}{\mu D^n}\left(\frac{\sigma}{N\phi(\alpha_1+\beta_1)} - 1\right), \\ R_i^* = \hat{R} - \eta(\theta_{YZ}, \Phi_{YZ}) + \sum_{t=i}^{YZ} \Delta_t. \end{cases} \tag{2.39}$$

Following (2.39), we are able to derive an optimal node coverage–contract reward pair upon the announcement of the UAV types. If the derived contract pairs satisfy the monotonicity conditions, they are the optimal contract formulation. Otherwise, we use the iterative adjusted algorithm, i.e., "Bunching and Ironing" algorithm [202], to return the results that satisfy the monotonicity constraint.

Note that to derive the optimal rewards for any type-i^* UAV, it is necessary to obtain the rewards for type-$i^* + 1, \ldots,$ type-YZ UAV. However, since there can only be one selected UAV for each subregion, the reward for the $(i^* + 1)$th UAV and onward has to be zero in practice. As such, to compute the rewards, we first assume the type-YZ UAV is selected to obtain a hypothetical R_{YZ}. With this, we can then work from backward, toward deriving $R_{YZ-1}, \ldots, R_{i^*}$. Thereafter, the model owner specifies route assignment $\mathcal{A}^{1,n}$ for the optimal UAV such that $\theta_1^{n*} = \frac{\sum_{i' \neq i, i' \in \mathcal{I}_n} a_{i,i'}^{1,n}}{|\mathcal{I}_n|}$.

In the next section, we consider the UAV-subregion assignment.

2.4 UAV-Subregion Assignment

In this section, we consider the matching-based UAV-subregion assignment using the GS algorithm. In Sect. 2.3, we note that the optimal UAV for each subregion has to be the lowest UAV type, i.e., type-1 UAV. However, given that all subregions will prefer type-1 UAV, there exists a need to consider a two-side matching such that the optimal UAVs are matched to the subregions efficiently.

2.4.1 Matching Rules

We introduce a complete, reflexive, and transitive binary preference relation [203], i.e., "$\succ$" to study the preferences of the UAV. For example, $n \succ_j n'$ implies that the UAV j strictly prefers subregion n to n', whereas $n \succeq_j n'$ indicates that the UAV j prefers the subregion n at least as much as UAV j'. We also consider the core definitions as follows:

Definition 2.3 *Matching:* For the formulated matching problem $(\mathcal{J}, \mathcal{N}, \mathcal{P}_n, \mathcal{P}_j)$, where $\mathcal{J}$ and $\mathcal{N}$ denote the set of UAVs and the set of subregions, respectively, whereas $\mathcal{P}_n$ and $\mathcal{P}_j$ denote the preferences of the subregions and UAVs, respectively. The matching $\mathcal{M}(j) = n$ indicates that the UAV j has been matched to subregion n, whereas the matching $\mathcal{M}(j) = \emptyset$ implies that the UAV j has not been matched to any subregion.

Definition 2.4 *Propose Rule:* The UAV $j \in \mathcal{J}$ announces its type to all the eligible subregions based on feasibility of task completion. Then, the contract is formulated and the subregion proposes to its most preferred UAV j^* in its preference set $\mathcal{P}_n$, i.e., $n^* \succ_j n^{*\prime}, n^* \forall n \in \mathcal{N}, n^* \neq n^{*\prime}$. Note that the preference of the subregion is managed by the FL model owner.

Definition 2.5 *Reject Rule:* The UAV $j \in \mathcal{J}$ rejects the subregion if a better matching candidate exists. Otherwise, the subregion that is not rejected will be retained as a matching candidate.

However, given that the subregion preference for UAV type is only based on two of the four type dimensions, i.e., marginal costs, some subregions may have multiple preferred UAVs with the same marginal costs. To that end, adopting the approach in [204], we introduce the rewards calibration rule by adjusting the traversal and transmission compensation downward to further reduce the number of eligible UAVs matched to each subregion.

Definition 2.6 *Rewards Calibration Rule:* For the subregion $n \in \mathcal{N}$ that has more than one optimal UAV matched, the contractual rewards can be adjusted downward, following which the preference of the UAVs is renewed for another iteration of matching. The adjustment is as follows:

$$\tilde{R}^n = \tilde{R}^n - \Delta \tilde{R}^n. \tag{2.40}$$

Algorithm 2.1 GS algorithm for UAV-subregion assignment

1: **Input:** $\mathcal{N}, \bar{\tau_n}, \mathcal{J}, \Psi, \mathcal{A}, \mathcal{B}, \mathcal{C}, E_j$
2: **Output:** $\mathcal{M}^*(j), \forall j \in \mathcal{J}$
3: Phase I: Initialization
4: **if** $\tau_P^{j,n}(\hat{\theta}) + \tau_C^{j,n}(\hat{\theta}) + \tau_T^{j,n} \leq \bar{\tau_n}(\hat{\theta})$ **then**
5:     UAV $j \in \mathcal{J}$ report types α_j^n, β_j^n to subregion n
6: **end if**
7: Sort $\Phi(\alpha_j^n, \beta_j^n)$ in an ascending order to derive $\mathcal{P}_n$
8: **Set** $\mathcal{R} = \mathcal{N}, \mathcal{M} = \emptyset$
9: Phase II: Iterative Matching
10: **while** $\mathcal{R} \neq \emptyset$ and $\mathcal{P}_n \neq \emptyset, \forall n \in \mathcal{R}$ **do**
11:     **for** n in $\mathcal{R}$ **do**
12:         $j^* = \arg\min_{\Phi_i \in \boldsymbol{\Phi}} \Phi_i$
13:         Formulate $(\theta_{j*}^n, R_{j*}^n)$ and propose contract
14:         **while** n has more than one optimal UAV **do**
15:             $\tilde{R}^n = \tilde{R}^n - \Delta\tilde{R}^n$
16:         **end while**
17:         **if** $u_{j*}^n(\theta_{j*}^n) > u_{j*}^{n'}(\theta_{j*}^{n'})$ **then**
18:             $\mathcal{M}(j^*) = n$
19:             Remove n from $\mathcal{R}$ and add n' to $\mathcal{R}$
20:         **else**
21:             Remove j^* from $\mathcal{P}_n$
22:         **end if**
23:     **end for**
24: **end while**

2.4.2 Matching Implementation and Algorithm

We now explain the implementation procedure of the UAV-subregion assignment.

Phase 1: Initialization

1. The model owner announces the sensing subregions and time constraint $\bar{\tau_n}$ for task completion to all UAVs in $\mathcal{J}$. Since θ_j^n is not known a priori, the time limit for task completion can be a function of the variable $\hat{\theta}$ specified by the model owner, e.g., $\hat{\theta} = 0.8$, i.e., only UAVs that can complete 80% of the route coverage should announce their types (Line 5 of Algorithm 2.1).
2. The UAVs announce their types α_j^n and β_j^n to the feasible subregions (Line 5), i.e., subregions in which the UAVs are able to meet the requirements for task completion.
3. We initialize (Lines 7–8):

    a. $\mathcal{M}$ as an empty set
    b. $\mathcal{R}$ as the set of subregions that have yet to be matched, i.e., $\mathcal{R} = \mathcal{N}$ at initialization
    c. $\mathcal{P}_n$ as the set of subregion preferences based on the ascending order of marginal costs $\Phi(\alpha_j^n, \beta_j^n)$ as discussed in Sect. 2.3.4.

Phase 2: Iterative Matching
Each iteration of matching consists of four stages.

1. *Proposal:* For each subregion $n \in \mathcal{R}$, the node coverage–contract reward pair is formulated for the most optimal UAV $j^* = \arg\min_{\Phi_i \in \boldsymbol{\Phi}} \Phi_i$ in the preference set $\mathcal{P}_n$. Then, the subregion proposes to an optimal UAV (Line 13).
2. *Rewards Calibration:* If the subregion has more than one optimal UAV, the contract reward is calibrated downward until a one-to-one matching is achieved (Line 15).
3. *Rejection:* If the UAV has a better matching candidate $u^n_{j*}(\theta^n_{j*}) < u^{n'}_{j*}(\theta^{n'}_{j*})$, the subregion is rejected. If not, the UAV keeps the subregion as a matching candidate.
4. *Update:* If a subregion has been matched, remove it from $\mathcal{R}$ (Line 19). If a prevailing matching candidate has been rejected, add it back to $\mathcal{R}$. Update $\mathcal{P}_n$ by removing the UAV that has issued a rejection in this iteration.

The iterations are repeated until all subregions have been matched, i.e., $\mathcal{R} = \emptyset$, or the remaining subregion has been rejected by all UAVs in its preference list, i.e., $\mathcal{P}_n = \emptyset$. The pseudocode is presented in Algorithm 2.1.

The stability and optimality properties of the GS algorithm are ensured following the proofs in [205]. As such, the self-revealing properties of our multi-dimensional contract design assure truthful type reporting, whereas the matching algorithm ensures that only one optimal UAV is matched to each region. In the following, we perform the performance evaluation of our incentive design.

2.5 Performance Evaluation

In this section, we consider the optimality of our devised contract. To illustrate the contract optimality, we first consider the case of six UAVs and a single subregion. Then, we study a single iteration of matching between 5 UAVs and 3 subregions. Finally, we study the GS matching-based UAV-subregion assignment that involves up to 7 UAVs and 6 subregions. Unless otherwise stated, the list of value ranges for the key simulation parameters is as summarized in Table 2.2. The key parameters we use are with reference to studies involving UAV and FL optimization [171, 197].

2.5.1 Contract Optimality

To illustrate the optimality of our multi-dimensional contract design, we first consider a highly simplified and demonstrative case of a single subregion and six UAVs of ascending marginal cost of route coverage. The values of α for the UAVs lie in the range of [250, 875], with increments of 125, whereas the β values lie in

Table 2.2 Table of key simulation parameters

Simulation parameters	Value
UAV sensing and traversal parameters	
p	10–35
v	10–20 m/s
l^n	1000–2000 m
$l^n_{C_i}$	500–1000 m
UAV computation parameters	
L	4
ε	$\frac{1}{3}$
δ	$\frac{1}{4}$
γ	2
A^*	0.6
ε	$\frac{1}{3}$
C	10–30 cycles/bit
K	24
V	4
κ	10^{-28}
f	2 GHz
UAV transmission parameters	
D^n	500–1000 MB
λ^n	10,000–15,000
H	1 MB
ρ	8–18
UAV and model owner utility parameters	
ϕ	0.05
μ	1
σ	100,000

the range of [20, 70], with increments of 10. The values are varied as presented in Table 2.2. Then, the auxiliary types are derived following (2.16) from Sect. 2.3.2 and are arranged in an ascending order for contract derivation. Type-1 UAV has the lowest marginal cost of node coverage, whereas type-6 UAV has the highest marginal cost of node coverage. To focus our study on the optimality of our contract design, we hold the traversal and transmission cost types of the UAV constant for now. In addition, we assume that all the UAVs can complete the task within the time constraints.

Figures 2.2 and 2.3 consider the hypothetical scenarios in which each particular UAV type takes turn to be matched to the subregion. As an illustration, if UAV type-1 has been matched to serve the subregion, the optimal node coverage is 1, whereas the contractual reward is 35. In contrast, if UAV type-6 is matched, the optimal node coverage is close to 0.4, whereas the reward is 20. From Figs. 2.2 and 2.3, we can observe that the monotonicity condition discussed in Theorem 2.1 from Sect. 2.3.3

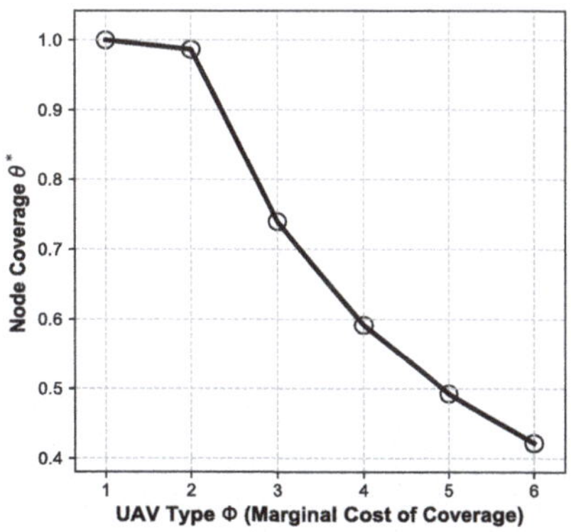

Fig. 2.2 UAV node coverage vs. auxiliary types

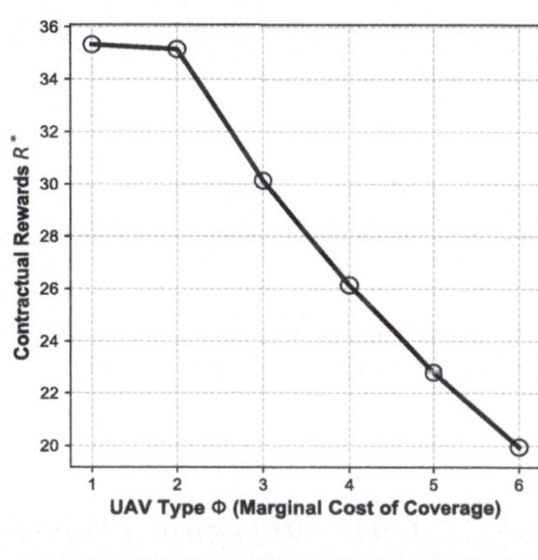

Fig. 2.3 Contract rewards vs. auxiliary types

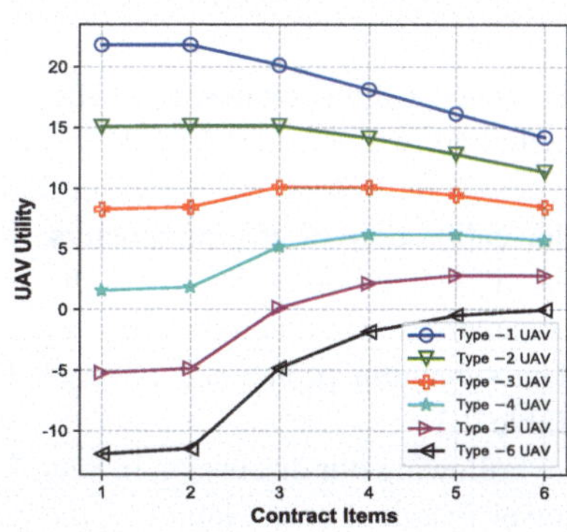

Fig. 2.4 Contract items vs. UAV utilities

holds. In other words, the higher is the marginal cost of node coverage, the lower is the optimal node coverage and contract rewards.

Figure 2.4 shows that the IC constraints of the contract hold. As an illustration, we consider the type-6 UAV, i.e., the UAV with the maximum marginal cost of node coverage. The type-6 UAV derives negative utility if it misreports its type, i.e., to imitate any other lower marginal cost UAV types 1–5. As discussed in Definition 2.2 from Sect. 2.3.1, each UAV derives the highest utility only if it reports its type

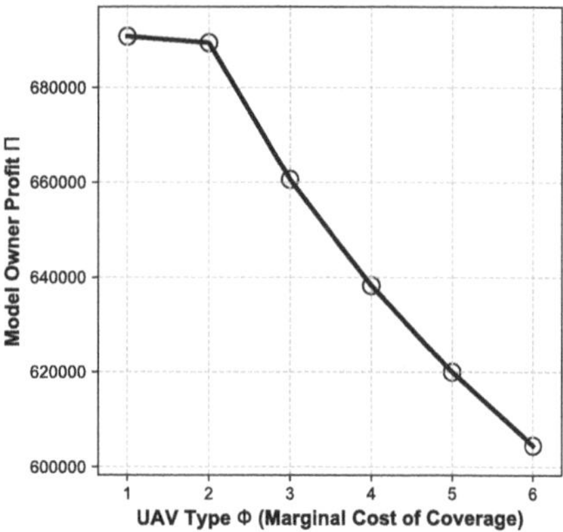

Fig. 2.5 The model owner profits vs. UAV auxiliary types

truthfully to the model owner. This validates the self-revealing mechanism of our contract.

Figure 2.5 shows that the model owner profits are the highest when it is able to be matched with the UAV of the lowest marginal cost of node coverage. This validates our discussion in Sect. 2.3.4 and confirms the model owner preference. In other words, among UAVs that are able to complete the task, the model owner prefers the UAV that incurs the lowest cost.

2.5.2 *UAV-Subregion Preference Analysis*

To analyze the preferences of the UAVs and subregions before we proceed with matching, we consider 5 UAV types and 2 subregions. In particular, the auxiliary types of the UAVs are shown in Table 2.3. Similarly, the types are derived from the calibration of the parameters listed in Table 2.3. The UAVs are sorted in the ascending order based on the marginal cost of node coverage. Besides, we consider three subregions 1, 2, 3 of coordinates $(1000, 1000)$, $(50, 50)$, $(500, 500)$. The subregion preference for each UAV is also presented in the last column of Table 2.3.

Following our discussion in Sect. 2.3.4, the subregions prefer the UAV types with lower marginal cost. Naturally, the preferences for all three subregions are similar as follows: $(1, 2, 3, 4, 5)$. Note that the subregions are indifferent between types 1 and 2, as well as types 4 and 5 given that the pairs have the same marginal cost of node coverage.

To consider the UAV preferences, we plot the potential profits that each UAV may gain from covering the different subregions in Fig. 2.6. Note that this profit is a hypothetical one in some cases since the profits can only be realized *if* the UAV has been matched to cover the subregion. However, given that the UAV is not aware if it will be matched to the subregion a priori, the preference list of the UAV can only be constructed with the assumption that it is indeed matched to the subregion. We

Table 2.3 UAV types for preference analysis

UAV type	Coordinates	α	β	Subregion preference
1	(100, 100)	500	20	(2, 3, 1)
2	(900, 900)	500	20	(1, 3, 2)
3	(400, 400)	750	30	(3, 2, 1)
4	(450, 450)	750	30	(3, 2, 1)
5	(500, 500)	1000	40	(3)

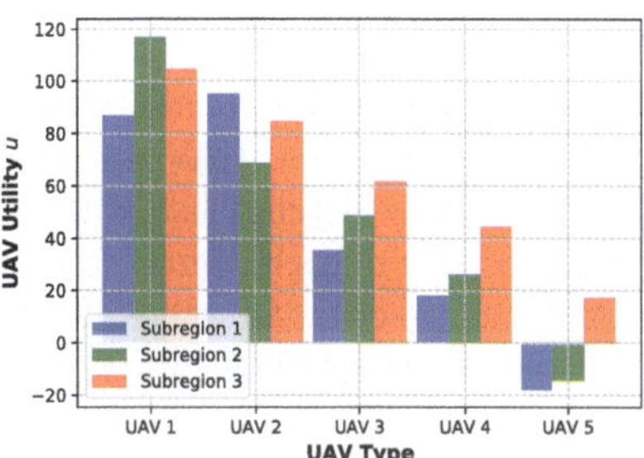

Fig. 2.6 The UAV utility for each subregion vs. types

note that UAV 1 prefers subregion 2, whereas UAV 2 prefers subregion 1, and so on. Intuitively, the preference for subregions relies on the traversal costs, i.e., the cost of traveling to and from the subregion. As such, the preferences for UAVs 1 and 2 are (2, 3, 1) and (1, 3, 2), respectively.

On the other hand, UAV 5 prefers only the closest region 3, given the potential negative profits derived if it serves the other two subregions, as a result of the high marginal costs incurred for task completion. As such, we are able to derive the matching of (Subregion 2, UAV 1) and (Subregion 1, UAV 2) given that the UAV-subregion preferences match perfectly.

The consideration for subregion 3 is clearly more challenging than that of subregions 1 and 2 given that the subregion is indifferent between the two remaining UAVs 3 and 4, and that the UAVs also rank the subregion highest, in terms of preference. To that end, we consider the rewards calibration rule proposed in Definition 2.6 from Sect. 2.4.1. The contract rewards are calibrated downward till a UAV emerges as the only choice left. In this case, after the downward calibration of rewards $\tilde{R}$, UAV 4 will clearly be matched with subregion 3, given its close proximity to the subregion.

Through this relatively straightforward example, we are able to derive an insight, i.e., a successful match will have the lowest marginal cost type UAVs matched to the subregion that it is situated closest to. Clearly, Fig. 2.6 also validates the efficiency of our incentive mechanism design, i.e., the best available UAV is matched to the respective subregion.

2.5.3 Matching-Based UAV-Subregion Assignment

In this section, we consider the matching-based UAV-subregion assignment. In particular, we consider three scenarios to illustrate the matching-based assignment.

In the *first* scenario, six UAVs of ascending marginal cost types are initialized to choose among six subregions that are of varying distances from each UAV. Each of the subregions is calibrated to hold the same quantities of data (D^n) and sensing area (l^n) for coverage. For ease of exposition, the UAVs are all able to complete their tasks within their energy capacities and stipulated time constraints. The coordinates of the subregions and UAVs, as well as the matching outcomes, are presented in Fig. 2.7. The preference list of the UAVs is presented in Table 2.4. Note that the preference list of each subregion is simply (1, 2, 3, 4, 5, 6), i.e., among all feasible UAVs that can cover the subregion within the time and energy constraints, the UAV with the lowest marginal cost is preferred.

From Fig. 2.7, we observe that the UAV 1 is matched to its most preferred subregion 6. Though UAV 2 also prefers subregion 6, it is unable to be matched to the subregion given that UAV 1 is higher up on the list of preferences of subregion 6. As such, UAV 2 is matched to its second choice. Naturally, the matching between UAV 3 and subregion 3, UAV 4 and subregion 2, as well as UAV 5 and subregion 5 is intuitive, given the unavailability of the other more preferred UAVs for the subregions to match with. We observe that UAV 6 is finally matched with its fifth choice, given that the UAV 6 has the lowest priority among subregions.

In the *second* scenario, we consider the same UAV types but with heterogeneous subregions of different data quantities and sensing areas for coverage. As is expected, the sizes of the regions do not affect the matching outcomes and the

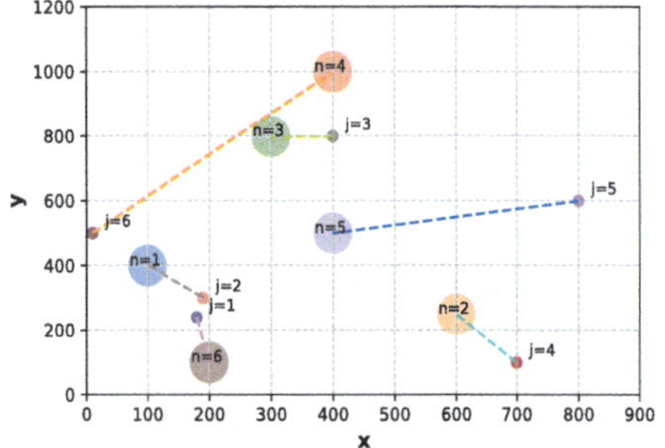

Fig. 2.7 UAV matching for homogeneous subregions

Table 2.4 UAV type and preference for subregions

UAV type	Subregion preference
1	(6, 1, 5, 2, 3, 4)
2	(6, 1, 5, 3, 2, 4)
3	(3, 4, 5, 1, 2, 6)
4	(2, 5, 6, 1, 3, 4)
5	(2, 5, 3, 4, 1, 6)
6	(1, 5, 3, 6, 4, 2)

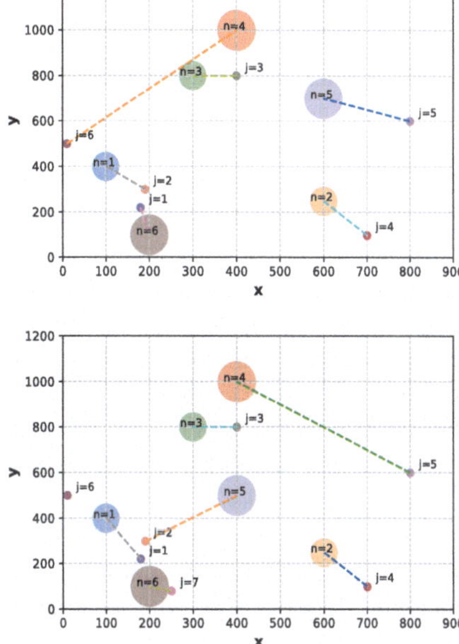

Fig. 2.8 UAV matching for subregions with different data quantities and coverage area

Fig. 2.9 UAV matching where $J > N$

matching remains the same (Fig. 2.8). This is given that the preference rankings of the subregion and the UAV remain constant. While the varying values of D^n and l^n affect the *magnitude* of UAV types, the *ordering* of the UAV types, and thus their preferences, is retained. This is important to ensure that the monotonicity of our contract design holds across subregions, so as to preserve the contract optimality.

In the *third* scenario, we consider the case where $J > N$, i.e., the number of UAVs exceeds the number of subregions available. As an illustration, we add in the UAV 7, which has the lowest marginal cost of node coverage relative to that of the other six available UAVs from the aforementioned scenarios. We observe from Fig. 2.9 that the matching outcomes have changed. UAV 7 is now matched with its most preferred subregion, i.e., subregion 6, in place of UAV 1. Naturally, this affects the assignment for the other UAVs. For example, UAV 1 has to be matched to its second choice now, whereas UAV 2 has to be matched to its third choice. We observe that the UAV of the largest type, i.e., UAV 6 is left out of the assignment as a result.

The simulation results allow us to validate the efficiency of our mechanism design. Firstly, the contract design ensures truthful type reporting and the incentive compatibility of our contract is validated. Secondly, with consideration of the preferences, the available UAV with the lowest marginal cost of node coverage is matched to the subregion. This ensures the profit maximization of the model owner.

2.6 Conclusion and Chapter Discussion

In this chapter, we have considered an FL based sensing and collaborative learning scheme involving UAVs for applications in the IoV paradigm. Given the incentive mismatches between the UAVs and the model owners, we have proposed a multi-dimensional contract matching incentive design such that the UAV with the lowest marginal cost of node coverage is assigned to each subregion for task completion.

As we have observed in this chapter, we have utilized the multi-dimensional contract optimization to account for various sources of heterogeneity in the network. While we have considered a three-dimensional contract in this chapter, it is possible to further extend this to account for other sources of heterogeneity. When utilized in the UAV network which is characterized by various operational complexities, the multi-dimensional contract has indeed proven to be a powerful tool for self-revelation in a network with information asymmetry. However, the chapter has the following drawbacks.

Firstly, even though UAVs have proven to be relatively useful and advantageous due to their flexibility in deployment and low cost, the reliability of UAVs is still uncertain. In particular, with weather uncertainties, the UAVs may fail to operate. However, in this chapter, we have not accounted for this uncertainty. One solution is to incorporate a backup UAV for the coverage of a region after the initially chosen UAV fails. The sensing and model training tasks of the UAVs can be conducted in multiple stages. After the first stage failure, a backup UAV can be deployed in the second stage and so forth. This step for the backup request can be added in Sect. 2.5.3, together with the matching phase. However, the backup UAV's addition implies additional cost incurred for the model owner. As such, to minimize the total network cost while accounting for the backup UAV reservation and corrective cost, we can utilize a *z-stage* stochastic optimization approach [206].

Secondly, in this chapter, we have assumed that the UAVs only train the FL model at their respective charging stations. For time-sensitive tasks, the delay incurred for the UAVs to return to their charging stations may be intolerable. In this chapter, we have omitted the discussion for brevity. Several works have introduced the concept of aerial computing, in which the UAVs perform model training onboard. However, with its limited battery constraint, this may be challenging. As a solution, the UAVs can also leverage offloading to edge servers to share this computation burden. To mitigate straggling edge servers, the work of [207] proposes a coded distributed computing approach in which a computation task can be partitioned into subtasks for offloading to the edge. Moreover, to preserve the privacy of FL, coding techniques such as random linear coding [208] can be utilized such that colluding edge servers are unable to recover the raw data of the UAVs.

Finally, in this chapter, we have assumed that the model owner is the sole model owner in the FL network. Following our discussion in Chaps. 1 and 2, there can be exceptions to this setting as follows: (a) the workers may be competing data owners who are reluctant to share their model parameters since the competitors also benefit from a trained global model and (b) the FL servers may compete with other FL servers, i.e., model owners. In this case, the formulation of the incentive mechanism design will be vastly different from that proposed. In Chap. 4, we will revisit this notion and devise a deep learning based auction mechanism in response, to model the competition among multiple model owners for the restricted participation of model owners.

Chapter 3
Joint Auction–Coalition Formation Framework for UAV-Assisted Communication-Efficient Federated Learning

3.1 Introduction

It is expected by 2023 that the IoT will have a market size of $1.1 trillion [209]. With an increasing number of vehicles being connected to the IoT, instead of just providing information to the drivers and uploading the sensor data to the Internet, from which other users can access the data, the vehicles are connected to each other in a network such that they are able to exchange their sensor data among each other for the optimization of a well-defined utility function. As a result, the deep integration between IoT and vehicular networks has led to the evolution of the Internet of Vehicles (IoV) [210] from the traditional Vehicular Ad Hoc Networks (VANETs).

Given the enormous amounts of dynamic data generated in the IoV network, Artificial Intelligence technologies are leveraged to build data-driven models. Specifically, many machine learning algorithms are developed to process and analyze data in order to make smart traffic predictions and provide intelligent route recommendations. To build the data-driven models, the traditional machine learning approaches require the data to be transmitted to the central data server for training. As the number of components connected to the IoV network increases, the amount of data generated in the IoV network increases exponentially. The central data server does not have sufficient storage and computing resources to store and process such massive amount of data, limiting the performance of the trained models. Besides that, the transmission of raw data over the communication channels incurs high communication cost and risks the privacy of the data as it may be eavesdropped while being transmitted.

In face of these challenges, Federated Learning (FL) [211] has gained great interest of the research and industrial communities. The IoV service provider runs an FL server, i.e., model owner, which employs multiple groups of parking vehicles or RSUs, i.e., workers, which collect information from their immediate surroundings. In each training iteration, the workers receive the model parameters from the model

W. Y. B. Lim et al., *Federated Learning Over Wireless Edge Networks*, Wireless Networks, https://doi.org/10.1007/978-3-031-07838-5_3

owner for model training on their respective local datasets. The updated local model parameters from all involved workers are then uploaded to the FL server which performs model aggregation. The model owner then transmits the updated model parameters to the workers for another iteration of local model training. A number of iterations are performed until a desired accuracy is achieved. Efficient data-driven models are built while preserving the privacy of the participating devices since only the model parameters, rather than the raw data, are exchanged.

However, as discussed in Chap. 1, communication inefficiency is a key bottleneck in the FL process. Since the federated network is connected to thousands of devices, it is estimated that the time taken to update the local parameters over the communication channels is longer than to compute the local parameters themselves. The data communication between the workers and the model owner may fail due to limited communication capacity and link failure [212, 213], which leads to the resulting node failure and missing node information. More than 30% of node failure contributes to a significant increase of the average end-to-end delay and concurrent communication time [214]. This means that the number of workers participating in the model training is reduced due to communication failure, thus causing a lower accuracy of the FL model.

Therefore, in this chapter, we propose the use of Unmanned Aerial Vehicles (UAVs) to facilitate the network in achieving higher communication efficiency in the FL network.

The vehicles and the RSUs may not have sufficient communication bandwidth and energy capacity to transmit the local model parameters directly to the FL server which is located far away. As a result, the FL server does not receive the local model parameters from these failing nodes, which cause a degraded performance in the trained FL model. The UAVs are increasingly used as relays between the ground base terminals and the network base stations [215] to improve network connectivity and extend coverage areas [216]. Since UAVs can be quickly deployed to designated areas whenever they are needed, the UAVs are nearer to the vehicles and the RSUs. Instead of transmitting the local model parameters to the FL server that is further away, the vehicles and the RSUs only need to transmit them to the UAVs that are located nearer to them, of which the parameters are relayed to the FL server. Hence, as compared to the direct communication between the IoV components and the FL server, the UAVs help to improve this communication by reducing node failures which may be due to limited resources and energy capacity of the IoV components.

In this chapter, we consider that the model owner is interested in facilitating FL among IoV components (Fig. 3.1), e.g., for traffic prediction. For efficient FL training, the model owner announces the FL task and the number of training iterations. The interested workers upload their sampling rate to the model owner. The model owner then selects its workers accordingly based on the importance of each worker, which is computed based on the uploaded sampling rates. Since the workers may not have sufficient resources, i.e., transmission power and communication bandwidth for efficient FL, the workers can utilize UAVs which have the required resources to facilitate the FL tasks. Firstly, the workers will conduct model training with their local datasets. Then, instead of sending the local model parameters to the remote FL

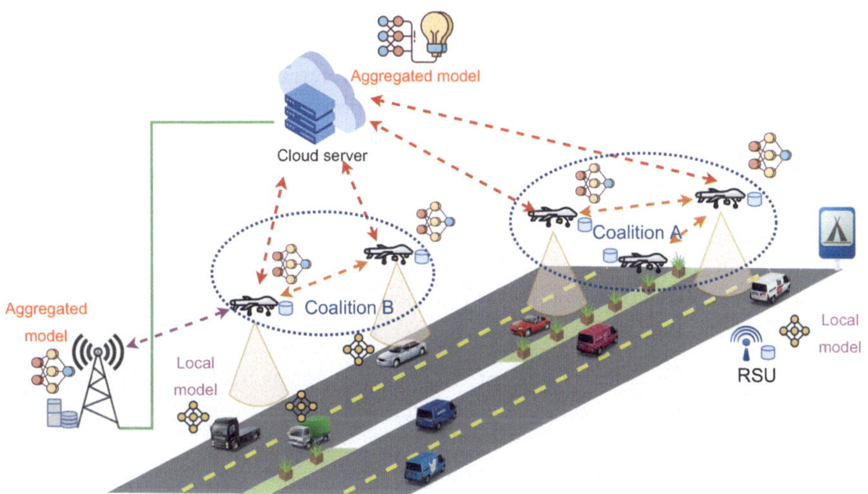

Fig. 3.1 System model consists of the cloud server (FL model owner), the vehicles and RSUs (selected FL workers), and the UAVs

server directly, the workers upload their local model parameters to the UAVs, which in turn aggregate the accumulated model parameters for relay to the model owner. Thereafter, the model owner aggregates the model parameters to derive the global FL model. The model owner will then announce the global FL model parameters to the workers for another round of FL training. The process continues until the number of iterations announced by the model owner is reached.

There are two advantages to this proposed approach. Firstly, the privacy of the workers is preserved since only local parameters of the workers are exchanged with the UAVs and the data of the workers remain locally at the workers' site. Secondly, with the help of UAVs, communication efficiency is increased, resulting in reduction of link and node failure.

However, a single UAV may not have sufficient energy to last for the entire training duration and other resources for the FL training process. As such, the UAVs may cooperate with other UAVs to ensure that the edge aggregation and communications with the model owner are provided throughout the entire FL training process. To model the cooperation among the UAVs, we propose a coalition formation approach among cooperating UAVs. In order to incentivize UAVs to participate in the FL training, the UAVs are rewarded for their contributions if they complete the FL training process. The UAVs receive different payoffs for supporting FL training process for different cells of workers as the importance of the workers varies. In addition, the workers have preference for different UAVs. In order to elicit the valuations of cells of workers for each coalition of profit-maximizing UAVs, we propose an auction scheme. Each cell of workers submit their bids to each coalition of UAVs. Collectively, the profit-maximizing UAVs evaluate all possible coalitional structure formation and decide on the optimal allocation of UAVs. Since the UAVs are profit maximizing, it is possible that there are two or more UAV coalitions which want to be allocated to the same cell of workers. In such a situation, based on the bids of the cells of workers, which also represent preferences of the cells of workers, the more preferred UAV coalition gets the allocation.

The main goal of this work is to develop a joint auction–coalition formation framework of UAVs toward enabling efficient FL for IoV networks. The integration of coalition formations of UAVs and optimal bidding of cells of workers makes the proposed framework suitable for the FL training process as it ensures that the UAVs have sufficient resources to complete the FL tasks and at the same time optimally allocates the profit-maximizing UAV coalitions to the groups of IoV components. Our key contributions include:

1. We propose the introduction of UAVs as wireless relays in FL network where communication efficiency between the model owner and the workers is improved.
2. We propose a joint auction–coalition formation framework where the problem of UAV coalition allocation to the groups of IoV components is formulated as a coalition formation game.
3. We propose a joint auction–coalition formation algorithm where a stable coalitional structure is achieved in which the allocation of profit-maximizing UAV coalitions to the groups of IoV components is solved by an auction scheme.

The remainder of the chapter is organized as follows. In Sect. 3.2, we present the system model and the problem formulation. Our proposed approach of coalition formation of UAVs and auction design are discussed in Sects. 3.3 and 3.4, respectively. Section 3.5 discusses the simulation results and analysis. Section 3.6 concludes the chapter.

3.2 System Model

We consider a UAV-assisted FL network over the IoV paradigm. The system model is comprised of an FL model owner which announces crowdsensing tasks to a group of selected FL workers. The FL workers that complete the crowdsensing tasks are responsible for labelling the data and training the model on their own local datasets. In order to perform data labelling for model training, the FL workers can leverage the knowledge, i.e., labels from a source domain party [217]. Thereafter, the local model parameters will be uploaded to the FL model owner for model aggregation. Since the communication efficiency between the model owner and the cell of workers is low due to link failure and missing nodes [212, 213], UAVs are introduced to facilitate the network to improve the communication efficiency of the FL model (Fig. 3.2).

We consider a coverage area that is divided into I cells, represented by a set $\mathcal{I} = \{1, \ldots, i, \ldots, I\}$. Each cell i contains W_i workers, where the total number of workers in each cell i is different and it is represented by

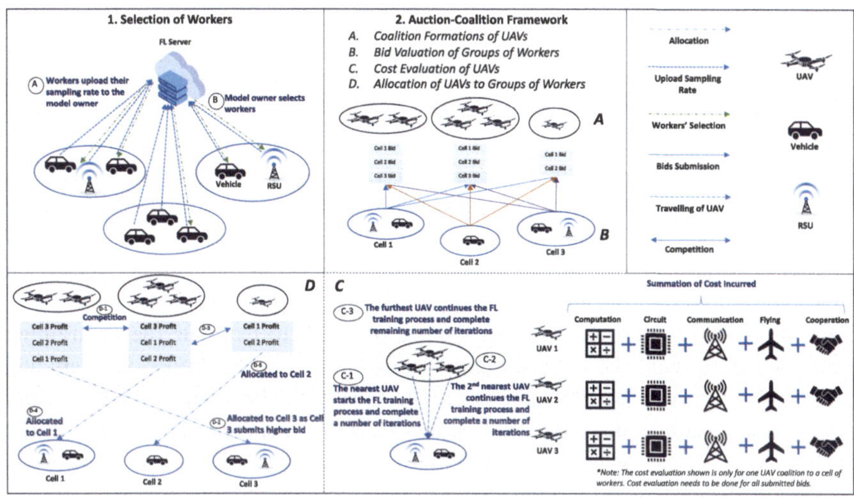

Fig. 3.2 Illustration of the joint auction–coalition formation framework

$\mathcal{W}_i = \{1, \ldots, w_i, \ldots, W_i\}$. The term d_{iw} denotes the wth IoV component in cell i. Each worker has different levels of importance, denoted by σ_{iw}, e.g., based on the sampling rate and the quality of data collected. The workers with low sampling rate and low quality are of low importance as there is insufficient data for the FL training process. The workers upload their sampling rates to the model owner, which in turn computes the importance of each worker. Since the design for incentive mechanisms that guarantee the truthfulness of the workers' sampling rates is not the focus of this chapter, existing incentive mechanisms such as the Vickrey–Clarke–Groves mechanism [218] can be implemented to ensure that the workers upload their true sampling rates. The incentive mechanisms are designed such that the players have no incentive to submit false information since their utilities decrease by doing so. Based on the importance level of each worker, the model owner selects the workers to participate in the FL training process. After selecting the workers, the model owner announces an FL model that will be trained over μ iterations. The optimal value of μ is determined such that the global loss function is minimized given the resource-constrained workers [219]. Upon completion of the FL task, the model owner will make a payment to the participating workers.

In the network, there are M geographically distributed UAVs, represented by a set $\mathcal{M} = \{1, \ldots, m, \ldots, M\}$ to facilitate the FL training process. The UAVs only collect the local model parameters from the individual workers after their model training and then perform edge aggregation before transmitting the aggregated local model parameters to the FL model owner. This ensures that the data belonging to the workers is not exchanged with any third parties, consequently preserving the data privacy. To improve communication efficiency, the UAVs need to ensure that they have sufficient energy to stay in the air throughout the entire FL process. In order to incentivize the UAVs to complete the FL training process, the workers will make a payment to the UAVs. In fact, the UAVs always maximize their payoffs. However, the individual UAVs may lack the energy capacities to stay in the air throughout the entire FL training process, and they cannot support certain cells of workers. As a result, the UAVs can consider to form coalitions. Note that each coalition of UAVs can only choose to facilitate the FL training process in one of the cells of workers. In order to elicit the willingness of payment of each cell of workers, an auction scheme is also explored to investigate the optimal allocation that maximizes the profits of the UAVs. To perform the auction process, the UAVs and the workers keep each other informed of their locations.

In our system model, we consider the use of UAVs to facilitate the FL training process by improving the communication efficiency between the FL workers and the FL model owner. At the same time, the UAVs can be used to support other types of users such as the typical mobile users, which generate background traffic for the UAVs. An extension to our system model to include different types of users is straightforward and can be considered in the future work. Note that with greater computational and communication capabilities to support more IoV components for the training of FL models with higher accuracy level, the operational cost of involving the UAVs in the model training may be higher than that without using the UAVs (Table 3.1).

3.2.1 Worker Selection

Each worker has different sensing capabilities and has different types of collected data. As such, the quality of data collected by each worker differs from one another. For example, a higher quality sensor provides data that more accurately reflects the changes in the road traffic environment. Reference [220] considers the quality of data collected as a function of the sampling rate. The quality of data implies the importance of the workers where a worker with higher quality of data has higher importance, such that the model accuracy can be improved. Following a similar assumption as in [220], the importance of worker w in cell i is denoted by

$$\sigma_{iw}(s_{iw}) = \frac{\log_{10}(s_{iw} + 1)}{\log_{10} 20}, \tag{3.1}$$

where s_{iw} is the sampling rate of a specific worker denoted by d_{iw}. This concave function implies that the increase in sampling rate at lower values has a greater impact on the quality of worker as compared to the same increase in sampling rate at higher values. The log term in Eq. (3.1) captures the effect of diminishing marginal rate of return on the sampling rate as the data may be replicated or similar, and thus, they are redundant [221].

The model owner selects workers which has an importance level greater than ζ, where ζ is the minimum threshold level of importance for the workers to be selected

Table 3.1 Table of commonly used notations

Parameter	Description
Worker parameters	
W_i	Total number of workers
s_{iw}	Sampling rate of worker
σ_{iw}	Importance of worker
Cell parameters	
I	Total number of cells
σ_i	Importance of cell
q_i	Price coefficient of importance of cell
UAVs parameters	
M	Total number of UAVs
d_i^m	Distance between UAV and cell
E_m	Energy capacity
K_m	Cooperation cost
δ_m	Price coefficient of energy cost of UAV
n_{im}	Maximum number of iterations of UAV
Coalitions parameters	
S_l	Coalition of UAVs
θ_1, θ_2	Weight parameters
$\alpha_i(S_l)$	True valuation
$x_i(S_l)$	Profit of coalition
$p_i(S_l)$	Cost of coalition
$c_i(S_l)$	Payment price of bid winner
$\gamma(\Pi)$	Profit of partition
Model owner parameters	
μ	Number of FL iterations

by the model owner. The value of ζ is determined by the model owner and is IoV application-specific. For example, ζ can be set larger for road safety applications such as navigation systems.

The importance of cell i can be computed as a summation of individual importance of each selected worker in cell i. It can be expressed as

$$\sigma_i = \sum_{w=1}^{W_i} \sigma_{iw}, \tag{3.2}$$

where σ_{iw} denotes the importance of worker d_{iw}.

3.2.2 UAV Energy Model

To execute the FL model training, each UAV incurs traveling cost to the workers' locations, communication, and computational costs. It also incurs coalition formation cost that includes cooperation and coordination cost if a UAV decides to join a coalition. The energy spent by the UAV in executing the task includes the following components:

- Flying from its original position to the destination and flying back to its original position
- Receiving local model parameters from the workers
- Staying at the destination to aggregate the local parameters from the workers
- Transmitting aggregated local model parameters to the model owner
- Receiving global model parameters from the model owner
- Transmitting global model parameters to the workers
- Receiving aggregated local model parameters from the workers
- Communicating with other UAVs to form a coalition

3.2.2.1 Flying Energy

Following a similar assumption as in [222], the energy required by the UAV m to fly from its depot to cell i can be denoted by $e^f_{i,m}$, where

$$e^f_{i,m} = \left(\frac{1}{2}g^f_i + \frac{1}{2}g^f_m\right) e^f \frac{d^m_i}{v}. \tag{3.3}$$

Without loss of generality, the UAVs are assumed to fly at a constant velocity v. The energy required to make a flight per unit time is denoted as e^f. We leverage g^f_i and g^f_m to denote the flying energy weight in cell i and the depot of UAV m, respectively. The flying energy weights depend on the terrain of the cells and the depot of the UAV, while d^m_i reflects the distance between UAV m and cell i. Note that the flying energy consumption, $e^f_{i,m}$, depends on both the locations of the UAV and the cell.

3.2.2.2 Computational Energy

Similar to [223], the energy needed by UAV m to complete the computation task per iteration, which is to aggregate the local model parameters of all workers, is represented as follows:

$$e^c_m = \kappa a_m f^2_m, \tag{3.4}$$

where a_m is the total number of CPU cycles needed by UAV m to complete the aggregation of local model parameters, f_m is the computation capability of UAV m which depends on the clock frequency of central processing unit (CPU) of UAV m, and κ is the coefficient of the value that depends on the circuit architecture of CPUs as in [224].

3.2.2.3 Communication Energy

For the model transmission between different entities in the system model, i.e., FL workers, the FL server, and the UAVs, an orthogonal frequency-division multiple access (OFDMA) is adopted.

The UAV requires energy to transmit data packets to the model owner. As in [225], the achievable uplink transmission rate by UAV m can be represented by

$$r_m^{UT} = \sum_{n=1}^{R} r_{mn} b_m \log_2 \left(1 + \frac{p_{mt} h_m}{I_n + b_m N_0}\right), \tag{3.5}$$

where R is the total number of resource blocks that the model owner allocates to the UAVs. Each UAV m is only allowed to occupy one resource block. The term r_{mn} is a resource block vector, where $r_{mn} = \{r_{m1}, \ldots, r_{mn}, \ldots, r_{mR}\}$. In particular, $r_{mn} = 1$ indicates that resource block n is allocated to UAV m, whereas $r_{mn} = 0$ represents that the resource is not allocated to UAV m. The term b_m is the bandwidth allocated to UAV m, p_{mt} is the transmission power of UAV m, h_m is the channel gain between UAV m and the model owner, I_n is the interference caused by other UAVs that uses the same resource block, and N_0 is the power spectral density of the Gaussian noise. Since the UAVs using the same resource block may interfere with each other and severely affect the uplink transmission rate between the UAVs and the model owner, interference needs to be considered. The issues of optimal resource block allocation, bandwidth allocation, and transmission power allocation have been extensively studied in the literature [226, 227]. Since it is not the focus of this chapter, existing highly efficient schemes such as an opportunistic and efficient resource block allocation (OEA) algorithm [226] can be applied to determine the allocation of resource blocks, bandwidth, and transmission powers of the UAVs. The energy required by UAV m to transmit model parameters in any cell i to the model owner over wireless channels per iteration is

$$e_m^{UT} = p_{mt} \frac{z^c}{r_m^{UT}}, \tag{3.6}$$

where z^c is the data size of cell-aggregated local model parameters. To upload the data packet of size z^c over time t, the data size that UAV m can handle must be larger than the input data size z^c.

The UAV also requires energy to receive the global FL parameters from the model owner. Similar to [97], the achievable downlink transmission rate by the FL server to UAV m is

$$r_m^{DR} = b_{BS} \log_2 \left(1 + \frac{h_m p_{BS}}{N_0 b_{BS}}\right). \tag{3.7}$$

The term of b_{BS} is the bandwidth that the model owner uses to broadcast information to the UAVs and p_{BS} is the transmission power of the model owner. The energy required by UAV m to receive data from the model owner over wireless channels per iteration is

$$e_m^{DR} = p_{mr} \frac{z_{BS}}{r_m^{DR}}, \tag{3.8}$$

where p_{mr} is the receiving power of UAV m. The term z_{BS} is the size of the global model parameters.

Similarly, energy is also needed by the UAVs to communicate with the workers in the cell. The achievable uplink transmission rate by worker w in cell i is

$$r_{iw}^{UR} = \sum_{n'=1}^{R_i'} r_{iwn} b_{iw} \log_2 \left(1 + \frac{p_{iw} h_{iw}}{I_{n'} + b_{iw} N_0}\right), \tag{3.9}$$

where R_i' is the total number of resource blocks allocated to worker w in cell i, and $r_{iwn'}$ is a resource block vector where $r_{iwn'} = \{r_{iw1}, \ldots, r_{iwn'}, \ldots, r_{iwR'}\}$. The bandwidth allocated to and transmission power of worker w in cell i are denoted by b_{iw} and p_{iw}, respectively. The term h_{iw} is the channel gain between the UAV and the worker and $I_{n'}$ is the interference caused by other workers using the same resource block.

The energy required by UAV m to receive local model parameters from all the workers in cell i per iteration is

$$e_{im}^{UR} = p_{mr} \sum_{w=1}^{W} \frac{z^w}{r_{iw}^{UR}}, \tag{3.10}$$

where z^w is the data size of the local model parameters of the worker.

The achievable downlink transmission rate by UAV m to the worker w in cell i is

$$r_{iw}^{DT} = b_m \log_2 \left(1 + \frac{h_{iw} p_{mt}}{N_0 b_m}\right). \tag{3.11}$$

The energy required by UAV m to transmit the global model parameters to all workers in cell i per iteration is

$$e_{im}^{DT} = p_{mt} \sum_{w=1}^{W} \frac{z_{BS}}{r_{iw}^{DT}}. \tag{3.12}$$

3.2.2.4 Hovering Energy

Hovering energy is the energy needed by the UAV to stay near the cell. The hovering energy of UAV m in any cell i is denoted by e_m^h.

3.2.2.5 Circuit Energy

Circuit energy is the energy consumed by the onboard circuits such as computational chips, rotors, and gyroscopes [228]. The onboard circuit energy is important and needs to be taken into consideration as it affects the energy efficiency of the network. The circuit energy of UAV m is denoted as e_m^t.

In the next section, we present the coalition formation game among the UAVs.

3.3 Coalitions of UAVs

In this section, we present coalitions of UAVs. In particular, each cell i charges the model owner for the local model parameters from the cells of workers. The amount of payment that the model owner needs to pay to the workers in cell i for their contribution to train the FL model depends on the importance of cell i, which is σ_i. In other words, the workers in cell i earn a revenue for training the FL model. The payment that cell i receives for participating in the FL training process is $\frac{q_i \sigma_i}{t^c}$, where q_i is the price paid by the model owner for a unit of importance in cell i and t^c is the total time needed to complete the FL task. Note that the model owner only makes payment to the workers upon completion of the FL task. Since the payment is dependent on the total time needed for the completion of the FL task, the workers have higher incentive to complete the FL task as fast as they can. Hence, the workers have higher incentive to bid for nearer UAV coalitions which is further explained in Sect. 3.4.

To receive full payment from the model owner, the cells need to complete μ iterations as required by the model owner. As a result, the cells of resource-constrained workers require the UAVs to facilitate the training process so as to improve communication efficiency. Since some UAVs may not have sufficient energy resources to support the entire FL process independently, the UAVs may form coalitions to support the cells of workers in completing the FL training.

3.3.1 Coalition Game Formulation

We have the following definitions for coalition game formulation.

Definition 3.1 A coalition of UAVs is denoted by $S_l \subseteq \mathcal{M}$, where l is the index of the coalition [229].

Definition 3.2 A partition or coalitional structure is a group of coalitions that spans all players in $\mathcal{M}$ and is represented as $\Pi = \{S_1, \ldots, S_l, \ldots, S_L\}$, where $S_l \cap S_{l'} = \emptyset$ for $l \neq l'$, $\bigcup_{l=1}^{L} S_l = \mathcal{M}$, and L is the total number of coalitions in partition Π.

The total number of possible partitions for M players is given by D_M, which is known as the Bellman number given as follows:

$$D_u = \sum_{j=0}^{u-1} \binom{u-1}{j} D_j, \text{ for } u \geq 1 \text{ and } D_0 = 1, \tag{3.13}$$

$\mathcal{M}$ denotes the coalition of all players, which is also known as the grand coalition.

Definition 3.3 A partition $\Pi = \{S_1, \ldots, S_l, \ldots, S_L\}$ is a stable partition if no coalition S_l has incentive to change the current partition Π by joining another coalition $S_{l'}$, where $S_l \cap S_{l'} = \emptyset$ for $l \neq l'$, or splitting into smaller disjoint coalitions [229].

The term of $K_m(S_l)$ is the cooperation cost incurred by UAV m to form a coalition S_l. It includes the cost of communicating with other UAVs in the same coalition. The cooperation cost is only incurred when there are two or more UAVs in a coalition. Otherwise, there is no cooperation cost incurred by the UAV. Specifically, K_m is defined as

$$K_m(S_l) = \begin{cases} K_m(S_l), & \text{if } |S_l| \geqslant 2 \\ 0, & \text{otherwise} \end{cases}, \ \forall S_l \in \Pi, \tag{3.14}$$

where $|S_l|$ is the number of coalition members in the coalition S_l. The coalition formation cost is a non-decreasing function of the size of the coalition. Note that there is no communication between the UAV coalitions.

When two or more UAVs cooperate to form a coalition S_l to support the FL training process in cell i, the FL training process starts as soon as the nearest UAV reaches cell i. This nearest UAV aggregates the local model parameters which are uploaded by the workers in cell i and transmits the aggregated local model parameters to the model owner for the maximum number of iterations that it can support given its energy capacity. Then, the second nearest UAV in the same coalition takes over, aggregates the local model parameters, and transmits the aggregated local model parameters to the model owner. Similarly, after the second nearest UAV completes its task, the third nearest UAV takes over and continues the process. If the maximum number of iterations that the UAV can support is greater than the remaining number of iterations of the FL training process, the UAV just needs to support the remaining number of iterations. If the required number of iterations announced by the model owner is completed by the nearer UAVs in a coalition, the farther UAVs do not need to support any iteration of the FL training process. For fairness purposes, since each UAV in the coalition completes a different number of iterations, the profit earned by each UAV in the coalition depends on the number of iterations it completes.

Given that each UAV m has an energy capacity of E_m, the maximum number of iterations that each UAV m in coalition S_l can support to facilitate the FL training process in cell i is denoted by $n_{im}(S_l)$, where

$$n_{im}(S_l) = \frac{E_m - 2e^f_{i,m} - K_m(S_l)}{e^{UR}_{im} + e^{DR}_m + e^{UT}_m + e^{DT}_{im} + e^c_m + e^h_m + e^t_m}. \tag{3.15}$$

Since each coalition S_l, $\forall S_l \in \Pi$ receives different revenues for facilitating the FL training process in different cell i, an auction scheme is needed to elicit the valuations of different cells for each coalition in order to decide on the optimal allocation of UAV coalitions to the cells of workers given a partition Π. The mechanism of the auction scheme is explained further in detail in the next section. Note that the total number of iterations that all UAVs in a coalition S_l, $\forall S_l \in \Pi$ can support determines the choices of coalitions that each cell of workers can bid for. The coalition formation game determines the stable partition that allows the UAVs to earn a maximum profit.

Proposition 3.1 *The formation of grand coalition, where all the UAVs join a single coalition, is not always stable.*

Proof In order to prove that the grand coalition is not always the optimal coalitional structure among the UAVs, we show that the coalitional game is not superadditive and prove that the core of the proposed game is empty by following the procedure in [230].

A coalitional game is superadditive if the cooperation among disjoint coalitions to form a larger coalition guarantees a payoff which is at least equal to that obtained by the disjoint coalitions separately, i.e., $x(S_1 \cup S_2) \geq x(S_1) + x(S_2)$, $S_1 \in \mathcal{M}$, $S_2 \in \mathcal{M}$ and $S_1 \cap S_2 = \emptyset$. Consider two coalitions where coalition S_1 is allocated to cell i to support the FL task in cell i and earning a profit of $x_i(S_1)$, whereas coalition S_2 is not allocated to any cell and earning zero profit, i.e., $x_i(S_2) = 0, \ \forall i \in \mathcal{I}$. By merging coalition S_1 and coalition S_2 to support the same cell i, the revenue, i.e., the payment price of cell i, does not change, but the total cost in terms of energy incurred and coalitional cost also increases. Thus, $x_i(S_1 \cup S_2) < x_i(S_1)$, i.e., the marginal profit of the UAVs from merging the two coalitions is negative. Therefore, the coalitional game is non-superadditive.

Algorithm 3.1 Algorithm for coalition formation of UAVs using merge and split

Input: Set of UAVs $\mathcal{M} = \{1, \ldots, m, \ldots, M\}$
Output: Partition that provides highest profit $\Pi^* = \{S_1^*, \ldots, S_y^*, \ldots, S_Y^*\}$

Initialize a set of partitions $\mathcal{Z} = \{\Pi_1, \ldots, \Pi_t, \ldots, \Pi_T\}$ that includes all possible partitions where T is the total number of partitions
Compute allocations of UAVs to cells of workers and the maximum obtainable profit given current partition, Π_{curr} from Algorithm 3.2
Merge Mechanism:
for each coalition S_l in Π_{curr} **do**
    Consider hypothetical partition Π_{new} where $S_l \cup S_{l'}, l \neq l'$
    Compute the optimal allocation of UAVs and maximum attainable profit, $\gamma(\Pi_{new})$ from Algorithm 3.2
    if $\gamma(\Pi_{new}) > \gamma(\Pi_{curr})$ **then**
        Merge coalitions S_l and $S_{l'}$
        Update partition $\Pi_{curr} \leftarrow \Pi_{new}$
        Update total profit $\gamma(\Pi_{curr}) \leftarrow \gamma(\Pi_{new})$
    end if
end for
return $\Pi_{merge} = \{S_1, \ldots, S_g, \ldots, S_G\}$
Split Mechanism:
for each coalition S_g in Π_{merge} **do**
    Initialize set of possible coalition splits $\tilde{S}_g = \{\tilde{S}_1, \ldots, \tilde{S}_g, \ldots, \tilde{S}_G\}$
    for each coalition $\tilde{S}_g$ in S_g **do**
        Consider hypothetical partition Π'_{new}
        Compute the optimal allocation of UAVs and maximum attainable profit, $\gamma(\Pi'_{new})$ from Algorithm 3.2
        if $\gamma(\Pi'_{new}) > \gamma(\Pi_{curr})$ **then**
            Split coalitions S_g such that $S_g \leftarrow \tilde{S}_g$
            Update partition $\Pi_{curr} \leftarrow \Pi'_{new}$
            Update total profit $\gamma(\Pi_{curr}) \leftarrow \gamma(\Pi'_{new})$
        end if
    end for
end for
return Final partition that provides the highest profit $\Pi^* = \{S_1^*, \ldots, S_y^*, \ldots, S_Y^*\}$

Next, we prove that the core of the proposed game is empty. As defined in [230], an imputation is a payoff vector $\mathbf{z} = \{z_1, \ldots, z_m, \ldots, z_M\}$ that satisfies the following two conditions:

1. Group rational, i.e., $\sum_{m \in \mathcal{M}} z_m = x(\mathcal{M})$
2. Individually rational where each player obtains a benefit no less than acting alone, i.e., $z_m \geq x(\{m\})\ \forall i$.

The core of the coalition is the set of stable imputations where there is no incentive for any coalition $S_l \in \mathcal{M}$ to reject the proposed payoff allocation $\mathbf{z}$, deviate from the grand coalition, and form coalition S_l. The core of the coalition is defined as

$$\tau = \left\{ \mathbf{z} \colon \sum_{m \in \mathcal{M}} z_m = x(\mathcal{M}) \text{ and } \sum_{m \in S_l} z_m \geq x(S_l)\ \forall S_l \in \mathcal{M} \right\}.$$

As we have previously established that the marginal profit of merging the coalitions can be negative, this means that the condition of individual rationality is violated, i.e., $\sum_{m \in S_l} z_m < x(S_l)$. In particular, the profit is higher if coalition S_1 does not merge with coalition S_2 to form a larger coalition. The core of the proposed coalitional game is empty.

Therefore, due to the non-superadditivity of the coalitional game and the emptiness of the core, the grand coalition does not form among the UAVs. Instead, smaller and disjoint coalitions of UAVs will form in the FL network. □

As such, we discuss the joint auction–coalition formation algorithm next.

3.3.2 Coalition Formation Algorithm

Following [231], we define two simple merge-and-split operations to modify a partition Π in $\mathcal{M}$.

- **Merge Rule:** Merge any set of coalitions $\{S_1, \ldots, S_l\}$ where $\gamma(\Pi) < \gamma(\Pi_{new})$, thus $\{S_1, \ldots, S_l\} \rightarrow \bigcup_{j=1}^{l} S_j$.
- **Split Rule:** Split any set of coalitions $\{\bigcup_{j=1}^{l} S_j\}$ where $\gamma(\Pi) < \gamma(\Pi'_{new})$, thus $\bigcup_{j=1}^{l} S_j \rightarrow \{S_1, \ldots, S_l\}$.

The merge-and-split algorithm for the coalition formation is presented in Algorithm 3.1.

The merge mechanism is illustrated from the perspective of a representative coalition S_l in a given partition Π. The coalition S_l decides to merge with another coalition $S_{l'}$ where $l \neq l'$ and $S_l \in \mathcal{M}$ if the total profit of the UAVs is improved. In order to evaluate the profits, the UAVs assume the hypothetical partition Π_{new} and compute maximum attainable profit of the hypothetical partition, $\gamma(\Pi_{new})$ (lines 5–6), of which the bidding of the workers and the allocation of UAVs are discussed in the next section. If the total profit of the UAVs in the hypothetical partition Π_{new}, which is denoted by $\gamma(\Pi_{new})$, is higher than that in its current partition Π_{curr}, which is represented as $\gamma(\Pi_{curr})$, coalition S_l will merge with coalition $S_{l'}$ to form new partition Π_{new} (lines 7–8). Then, given this new partition Π_{new}, the current partition and the value of the maximum profit will be updated (lines 9–10). On the other hand, $\gamma(\Pi_{new})$ is less than that of the current partition Π_{curr}, $\gamma(\Pi_{curr})$, and there will be no change to the current partition Π_{curr}. For the next iteration, any coalition S_l in the given partition, Π_{curr}, will consider to form a coalition with another coalition $S_{l'}$. The merge mechanism terminates when all possible partitions have been considered. At the end of the merge mechanism, the algorithm returns a resulting partition $\Pi_{merge} = \{S_1, \ldots, S_g, \ldots, S_G\}$ (line 13). Consequently, the split mechanism is carried out.

Similarly, the split mechanism is illustrated from the perspective of a representative coalition S_g in partition Π_{merge}. Let $\tilde{S}_g$ be the set of possible coalition splits for coalition S_g, where $\tilde{\mathcal{S}}_g = \{\tilde{S}_1, \ldots, \tilde{S}_g, \ldots, \tilde{S}_G\}$ (line 16). Coalition S_g considers one of the possible ways to split its coalition into smaller disjoint coalitions (line 17). Similar to the merge mechanism, in order to evaluate the profits, the UAVs assume the hypothetical partition Π'_{new} and retrieve the bids from the auction and compute their maximum obtainable profits (lines 18-19). If the total profit of the UAVs is improved with more disjoint coalitions, the coalition S_g will proceed to split and form a new partition Π'_{new} (lines 20-21). The current partition and the value of maximum profit will be updated (lines 22-23). Then, given the new partition Π'_{new}, any coalition will split and evaluate their profits again. The partition is maintained if the profit is higher than if the split is carried out. The split mechanism terminates when all possible partitions with smaller coalitions have been considered. At the end of the merge-and-split algorithm, the final partition $\Pi^* = \{S_1^*, \ldots, S_y^*, \ldots, S_Y^*\}$ is returned (line 27).

Proposition 3.2 *The final partition* $\Pi^* = \{S_1^*, \ldots, S_y^*, \ldots, S_Y^*\}$ *is a stable partition that maximizes the total profit of the UAVs.*

Proof Given any current partition $\Pi_{curr} = \{S_1, \ldots, S_l, \ldots, S_L\}$, where $S_l \cap S_{l'} = \emptyset$ for $l \neq l'$, $\bigcup_{l=1}^{L} S_l = \mathcal{M}$, the current partition Π_{curr} is only updated when the total profit of the hypothetical partition Π_{new} is higher than that of the current partition Π_{curr} by either merging two disjoint coalitions or splitting a single coalition into multiple smaller disjoint coalitions. Therefore, the merge-and-split algorithm (Algorithm 3.1) generates a sequence of partitions where each partition has either equal profit as or higher profit than the partition from previous iteration. The total number of possible partition D_M is a finite number. The algorithm will eventually terminate at a partition $\Pi^* = \{S_1^*, \ldots, S_y^*, \ldots, S_Y^*\}$ where the total profit of partition Π^* cannot be increased further by merging or splitting coalitions. Hence, the final partition $\Pi^* = \{S_1^*, \ldots, S_y^*, \ldots, S_Y^*\}$ is stable. In addition, at stable partition $\Pi^* = \{S_1^*, \ldots, S_y^*, \ldots, S_Y^*\}$, since the total profit of the UAVs cannot be increased further by changing the partition Π^*, the total profit given by partition $\Pi^* = \{S_1^*, \ldots, S_y^*, \ldots, S_Y^*\}$ is the maximum. □

Next, we present an auction design to elicit the valuations of cells of workers for all possible UAV coalitions in order to determine the allocation of UAV coalitions to the cells of workers, thereby maximizing the profit of the UAVs given a partition Π.

3.4 Auction Design

We consider multiple auctioneers and multiple buyers at the same time. The UAVs are the auctioneers which conduct the auctions and offer the resources to facilitate the FL training process. The cells of workers are the buyers that submit bids and pay the UAVs for their energy resource.

3.4.1 Buyers' Bids

Firstly, the workers in cell i ($\forall i \in \mathcal{I}$) submit their bids to each possible coalition S_l in a given partition Π to indicate their valuations for each coalition.

Each cell needs to build their own preference over all the possible coalitions that can provide services, where cells of workers need to compare the coalitions and rank their preferences. As such, the concept of preference relation is introduced to evaluate the order of preference of each player.

Definition 3.4 For any player $i \in \mathcal{I}$, a preference relation [232] $\succeq_i$ is defined as a complete, reflexive, and transitive binary relation over the set of all coalitions that cell i can choose.

In particular, for any cell $i \in \mathcal{I}$, given $S_1 \subseteq \mathcal{M}$ and $S_2 \subseteq \mathcal{M}$, $S_1 \succeq_i S_2$ means that cell i prefers coalition S_1 over coalition S_2, or at least cell i prefers both coalitions equally. The asymmetric counterpart of $\succeq_i$, which is represented by $\succ_i$, when used in $S_1 \succ_i S_2$, means that cell i strictly prefers coalition S_1 over coalition S_2. It is worth noting that the preference relation is defined to quantify the preferences of the players, which can be application-specific. For example, the IoV components prefer UAVs that are nearer for road safety applications but prefer UAVs that provide higher bandwidth for entertainment services. It can represent a function of parameters such as the utility gain by joining the coalition.

The preference relation for any cell $i \in \mathcal{I}$ can be represented by the following formula:

$$S_1 \succeq_i S_2 \iff \alpha_i(S_1) \geq \alpha_i(S_2), \tag{3.16}$$

where $S_1 \subseteq \mathcal{M}$ and $S_2 \subseteq \mathcal{M}$, $\alpha_i(S_l)$ is the preference function for any cell $i \in \mathcal{I}$ and for any coalition S_l facilitating FL training process in cell i.

In general, the more important cells of workers have incentives to pay the UAVs more to support the FL training process as they receive higher reward for completing the FL task. Meanwhile, the valuations of the cells for the UAVs depend on the latency of the task, i.e., how quickly the model parameters are transmitted to the model owner. As such, the cells of workers have more incentive to employ UAVs that are nearer to them since shorter time is required for them to reach the cell to facilitate FL training. Although the FL training process starts as soon as the nearest UAV in a coalition reaches the cell of workers, the valuation of the cell of workers for the UAV coalition does not depend on the nearest UAV in the coalition. In fact, the valuation of the cell of workers for a coalition S_l depends on the farthest UAV in the coalition as they need to wait for the farthest UAV to arrive to continue and complete the FL training process, even if other nearer UAVs can reach the cell and support their parts of the training process in a short amount of time. Specifically, the time for the farthest UAV to reach the cell may be longer than the time needed by the nearer UAVs to travel to the cell and complete their parts of the FL training process.

Thus the valuation of workers in cell i for coalition S_l, which is also the preference function, is defined as

$$\alpha_i(S_l) = \theta_1 q_i \sigma_i + \frac{\theta_2}{\max_{m \in S_l} t_i^m}, \tag{3.17}$$

where t_i^m is the time needed for UAV m to travel to cell i, θ_1 is the weight parameter of the importance of the cell of workers, whereas θ_2 is the weight parameter of the time needed for the UAVs to reach the cells. In particular, each cell i submits its valuations for each UAV coalition in a given partition Π. Each cell i will only submit its bids to any coalition S_l, $\forall S_l \in \Pi$ if the total number of iterations that can be completed by the coalition S_l is greater than the required number of iterations specified by the model owner. In other words, coalition S_l will only receive bids from any cell i if the following condition is satisfied:

$$\sum_{m \in S_l} n_{im}(S_l) \geq \mu. \tag{3.18}$$

The traditional first-price auction, in which the highest bidder pays the exact bid it submits, maximizes the revenue of the seller but does not guarantee that the bidders submit their true valuations. In other words, the bidders have incentives to not submit their true valuations. In particular, if the bidder bids lower and wins the bid, its utility increases. Hence, by adopting the first-price auction, the property of incentive compatibility of the auction is not guaranteed. As such, the second-price auction is adopted to determine the payment price of the bid winners. The payment price is determined to incentivize the cells of workers to honestly report their true valuations such that the valuations of the cells of workers are equal to their bids, i.e., $\alpha_i(S_l) = \lambda_i(S_l)$, where $\lambda_i(S_l)$ is the bid submitted by cell i to coalition S_l. The payment price of cell i for coalition S_l is represented by $p_i(S_l)$. Thus, the utility of cell i if it wins the bid is defined as

$$u_i(S_l) = \alpha_i(S_l) - p_i(S_l). \tag{3.19}$$

On the other hand, if cell i is not allocated any UAV coalition, the utility is $u_i(S_l) = 0$.

3.4.2 Sellers' Problem

The UAVs as the auctioneers need to decide on the optimal allocation to different cells of workers. The algorithm for the allocation of profit-maximizing UAVs is presented in Algorithm 3.2.

The objective of the UAVs is to maximize sum of their individual profits where the UAVs cooperate fully with each other and are not selfish. In order to calculate the maximum total profits for any partition Π, the maximum possible profit for each coalition in the partition needs to be calculated first. The coalition will only earn the actual profit if it is allocated to support the FL task in any cell i. Given any partition Π, the revenue of coalition S_l in supporting cell i is the payment price, i.e., $p_i(S_l)$, which can be evaluated from the bid values (lines 3–4). The cost of coalition S_l in supporting cell i is the summation of energy incurred by each UAV in the coalition (line 5). The cost of coalition S_l in supporting cell i is denoted as $c_i(S_l)$, which is defined as

$$c_i(S_l) = \delta_m \sum_{m \in S_l} 2e^f_{i,m} + \sum_{m \in S_l} K_m(S_l) + \delta_m \sum_{m \in S_l} n^c_{im}(e^{UR}_{im} + e^{DR}_m + e^{UT}_m + e^{DT}_{im} + e^c_m + e^h_m + e^t_m), \tag{3.20}$$

where n^c_{im} is the number of iterations that is supported by UAV m in cell i and δ_m is the cost coefficient per unit of energy spent by UAV m. Note that it is not necessary that each UAV in the coalition supports the maximum number of iterations, n_{im}. If the number of remaining iterations is less than the maximum number of iterations that the UAV can support, the UAV needs to only support the remaining number of iterations, which requires less energy.

The possible profit earned by coalition S_l for supporting FL training process in any cells $i \in \mathcal{I}$ is the difference between the revenue earned and the cost incurred by coalition S_l in facilitating the FL training in cell i (line 5). The possible profit denoted by $x_i(S_l)$ can be represented by

$$x_i(S_l) = p_i(S_l) - c_i(S_l). \tag{3.21}$$

In this regard, each coalition S_l in partition Π can now evaluate the most preferable cell that provides them with the maximum profit (line 12). If two or more coalitions choose the same cell of workers, the coalition of UAVs which is the most preferred based on the bids submitted will be allocated (lines 13–14). For example, according to the preference relation in Equation (3.16), coalition S_l will be allocated to cell i, instead of coalition $S_{l'}$ if $\alpha_i(S_l) \geq \alpha_i(S_{l'})$. The coalition of UAVs which are not allocated any cell will not earn any profit. If there is no competition among coalitions of UAVs for a cell, the UAV coalition is allocated to the cell (lines 15–16). Given partition Π and the allocation of UAV coalitions to the cells, the total profits can be calculated by summing up the individual profits of each coalition which is allocated a cell (lines 18–19). The total profit of partition Π is given by $\gamma(\Pi)$, which is defined as follows:

$$\gamma(\Pi) = \sum_{i=1}^{I} \sum_{l=1}^{L} \beta_{il} x_i(S_l), \tag{3.22}$$

Algorithm 3.2 Algorithm for allocation of profit-maximizing UAVs

Input: Current partition $\Pi = \{S_1, \ldots, S_l, \ldots, S_L\}$
Output: Maximum obtainable profit, $\gamma(\Pi)$ and allocation of UAVs to cells of workers
1: **for each** cell of workers i in $\mathcal{I}$ **do**
2:     **for each** coalition of UAVs S_l in Π **do**
3:         Submit bid values, $\alpha_i(S_l)$
4:         Compute payment price, $p_i(S_l)$ if cell i is the bid winner for coalition S_l
5:         Compute the cost of each UAV coalition S_l in supporting cell i
6:         Compute the profit of each UAV coalition S_l in supporting cell i, $x_i(S_l)$
7:     **end for**
8: **end for**
9: Initialize $t = I$
10: Initialize $\gamma(\Pi) = 0$
11: **while** $t \neq 0$ **do**
12:     Identify the highest profit for each coalition and its corresponding cell, $x_i(S_l)$ in Π
13:     **if** Two or more UAV coalitions compete for the same cell **then**
14:         The cell with the highest bid is allocated the coalition
15:     **else**
16:         The coalition (without competition) is allocated the cell
17:     **end if**
18:     Update partition Π
19:     $\gamma(\Pi) \leftarrow \gamma(\Pi) + x_i(S_l)$
20:     Remove allocated UAVs and cell from the partition Π
21:     $t = t - 1$
22: **end while**
23: **return** $\gamma(\Pi)$ and allocation of UAVs to cells of workers

where $\beta_{il} = \{\beta_{i1}, \ldots, \beta_{il}, \ldots, \beta_{iL}\}$ is an allocation vector which indicates whether coalition S_l is allocated to cell i. In particular, $\beta_{il} = 1$ if coalition S_l is allocated to cell i and $\beta_{il} = 0$ if coalition S_l is not allocated to cell i.

Then, the allocated coalition is removed from the current partition, Π, and the corresponding cell is also removed from the list of cells (line 20). The number of cells to be supported reduces by one (line 21).

3.4.3 *Analysis of the Auction*

We now analyze the individual rationality and the incentive compatibility of the auction mechanism. Firstly, to encourage the buyers to participate in the auction, the auctioneers need to ensure that the buyers receive positive payoffs. Moreover, the auction mechanism discourages the buyers from submitting untruthful valuations that do not reflect the true valuations of buyers' bids.

Proposition 3.3 *Each buyer, i.e., cell of workers, achieves individual rationality in this auction mechanism.*

Proof Firstly, for each cell of workers which has no winning bid, its utility is zero, i.e., $u_i(S_l) = 0$ since it does not earn a revenue from the model owner and it also does not pay any UAV coalition to complete the FL task. Secondly, for the cells of workers with winning bids, i.e., allocated a UAV coalition to support the FL training process, their utility is

$$u_i(S_l) = \alpha_i(S_l) - p_i(S_l) = \lambda_i(S_l) - p_i(S_l). \tag{3.23}$$

The utility gained by the bid winner is non-negative, i.e., $u_i(S_l) \geqslant 0$. When $\lambda_i(S_l)$ wins the bid, its true valuation must be greater than the payment price, i.e., $\alpha_i(S_l) \geqslant p_i(S_l) = \lambda_{i'}(S_l)$. Otherwise, $\lambda_{i'}(S_l)$ wins the bid instead. □

In order to prove the property of incentive compatibility, i.e., the truthfulness of buyers' bids, we show that there is no incentive for the buyers to submit bids other than their true valuations by following the procedure in [233]. The auction mechanism is truthful if it satisfies the following two conditions:

1. The winning bid algorithm is monotonic.
2. The pricing is independent of the winning bid.

Proposition 3.4 *The winning bid algorithm is monotonic. For each bid $\lambda_i(S_l)$, if bid $\lambda_i(S_l)$ wins, then bid $\lambda_i'(S_l)$ also wins, where $\lambda_i'(S_l) = \lambda_i(S_l) + \phi$ and $\phi > 0$.*

Proof Suppose that $\lambda_i(S_l)$ is the winning bid with the UAV coalition S_l earning a profit $x_i(S_l)$, a higher bid $\lambda_i'(S_l)$, where $\lambda_i'(S_l) = \lambda_i(S_l) + \phi$ and $\phi > 0$, will result in the same profit $x_i(S_l)$, since the cost of coalition S_l for supporting the FL training process in cell i and the price of the winning bid $p_i(S_l)$ are the same, $\lambda_i'(S_l)$ still wins the bid. □

Proposition 3.5 *If cell i wins the bidding with $\lambda_i(S_l)$ and $\lambda_i'(S_l)$, the payment price $p_i(S_l)$ is the same for both bids.*

Proof As long as cell i wins the bid, the payment price of the bid winner is the second price. By bidding either $\lambda_i(S_l)$ or $\lambda_i'(S_l)$, the payment price does not change. Hence, the payment price is independent of the winning bid $\lambda_i(S_l)$ or $\lambda_i'(S_l)$. □

Proposition 3.6 *No buyer, i.e., cell of workers, can achieve higher payoff by bidding with values other than its true valuations, i.e., if a buyer bids $\lambda_i'(S_l) \neq \lambda_i(S_l)$, its utility is $u_i'(S_l) \leq u_i(S_l)$.*

Proof Following a similar procedure in [233], we consider two cases.

Case 1. If cell i wins the bid for coalition S_l, it needs to pay $p_i(S_l)$ which is equal to the second highest winning bid. If cell i decides to bid untruthfully with $\lambda_i'(S_l) \geq \lambda_i(S_l)$, cell i still wins the bid according to Proposition 3.4 without any increase in utility. On the other hand, if cell i decides to bid untruthfully with $\lambda_i'(S_l) \leq \lambda_i(S_l)$, cell i may still win the bid without any increase in utility according to Proposition 3.5 or lose the bid, which leads to a reduction in utility.

Case 2. If cell i loses the bid for coalition S_l, it does not need to pay anything. If cell i bids untruthfully with $\lambda_i'(S_l) \geq \lambda_i(S_l)$, cell i may win the bid but suffer from a non-positive utility or lose the bid with no change in utility. If cell i bids untruthfully with $\lambda_i(S_l) \leq \lambda_i(S_l)$, it will still lose the bid. Therefore, the buyer is not incentivized to bid untruthfully since by doing so, it is not able to achieve higher utility.

□

Therefore, the auction mechanism has the properties of individual rationality and incentive compatibility.

3.4.4 Complexity of the Joint Auction–Coalition Algorithm

In this subsection, we investigate the complexity of the joint auction–coalition algorithm. The time complexity of the proposed algorithm depends on the number of attempts of the merge-and-split operations. Since for each attempt involves the auction-based allocation of which the time complexity increases linearly with the number of cells of workers, the overall time complexity of the proposed algorithm is given by the multiplication of the number of merge-and-split attempts and the complexity of the auction-based allocation.

For a given partition, the merge operation is carried out if the merging of two coalitions results in a higher total profit than if the two coalitions operate independently. In the worst case scenario, each UAV coalition attempts to merge with all other UAV coalitions. At the start, all UAVs form singleton coalitions, which result in M coalitions. In the worst case, $\frac{M(M-1)}{2}$ number of attempts required before the first merge operation is carried out, $\frac{(M-1)(M-2)}{2}$ number of attempts required before the second merge operation is carried out, and so on. The total number of attempts of merge operations is in $O(M^3)$ for the worst case scenario. However, in practice, the merge process requires a significantly lower number of attempts. Once the UAV coalitions merge to form a larger coalition, the number of merge attempts per coalition reduces since the number of merging possibilities for the remaining UAVs decreases.

In the worst case scenario, splitting a UAV coalition S_l involves finding all possible partitions of the UAVs in the coalition. The number of possible partitions of a set of UAVs is given by the Bellman number in Eq. (3.13), which increases exponentially in the number of UAVs in the set. In practice, the split operation is not performed over the set that consists of all UAVs, but over the smaller set of each formed UAV coalition. Since the UAVs incur higher cooperation cost by forming larger coalitions in Eq. (3.14), they prefer to form coalitions of smaller size where possible. As a result, the split operation is affordable in terms of complexity.

3.5 Simulation Results and Analysis

In this section, we evaluate the joint auction–coalition formation framework of the UAVs. Firstly, we evaluate communication efficiency in the UAV-enabled network. Then, we analyze the preference of each cell for each UAV, followed by the profit-maximizing behavior of the UAVs and the allocation of UAVs to different cells of workers. Finally, we compared the proposed framework against existing schemes. We consider a distributed FL network on a Cartesian grid of size 1000×1000 as shown in Fig. 3.3. All simulation parameters are summarized in Table 3.2.

The sensors of the IoV components such as the vehicles and the RSUs typically have sampling rates that range between 250 samples per second (sps) and 32000 sps. The IoV components, i.e., the FL workers, are selected to train the FL model if their importance, $\sigma_{iw} \geq 1$. The training datasets that the FL workers use to train the FL model consist of the samples collected by their sensors. Thus, the FL workers with sensors of higher sampling rates have larger training datasets as compared to the FL workers with sensors of lower sampling rates. Note that since the selected FL workers only transmit the local model parameters to the model owner, the size of the training datasets does not affect the size of the transmitted local model parameters.

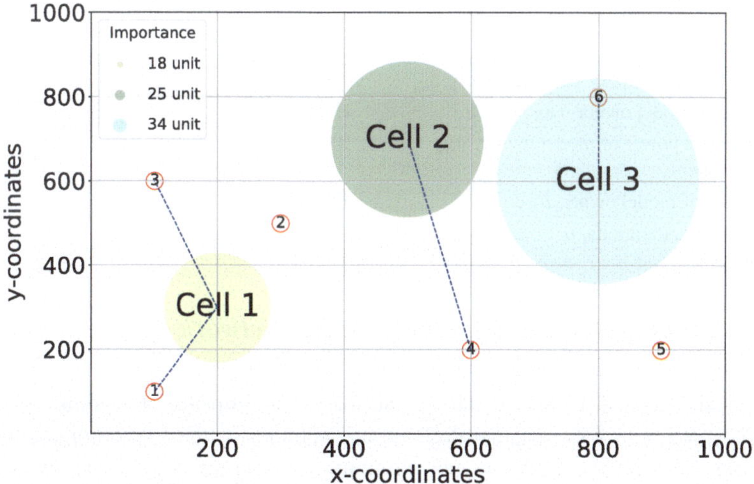

Fig. 3.3 Distributed FL Network with 3 cells and 6 UAVs

Table 3.2 Simulation parameters

Parameter	Values
Total number of cells, I	3
Total number of UAVs, M	6
Minimum threshold level of importance, ζ	1
Sampling rate of worker, s_{iw}	250 sps–32,000 sps
Bandwidth of worker, b_{iw} [225]	50–150 kHz
Transmission power of worker, p_{iw} [225]	1–10 mW
Channel gain of worker, h_{iw}	2–8 dBm
Data size of worker, z^w	100 MB
Data size of cell, z^c	500 MB
Flying energy per unit time, e^f	1000 mW
Flying velocity of UAV, v	10 m/s
Computational coefficient, κ [223]	10^{-26}
Computational capability of UAV, f_m	10^8
Total number of CPU cycles, a_m	1–3 GHz
Hovering energy, e_m^h	5 J
Circuit energy, e_m^t	5 J
Bandwidth of UAV, b_m	200–400 kHz
Transmission power of UAV, p_{mt} [234]	0.5–5 W
Receiving power of UAV, p_{mr}	0.5–1 W
Channel gain of UAV, h_m	5–25 dBm
Cooperation cost, K_m	2
Price coefficient of UAV per unit energy, δ_m	0.03
Data size of model owner, z_{BS}	1500 MB
Transmission power of model owner, p_{BS} [234]	10 W
Bandwidth of model owner, b_{BS}	5000 kHz
Number of FL iterations, μ	20

3.5.1 *Communication Efficiency in FL Network*

The introduction of UAVs to the FL network can improve the communication efficiency. Without loss of generality, we use vehicles as the IoV components that communicate with the UAVs in our simulations. According to [225], we set the bandwidth of IoV vehicles, denoted by $b_{iw} = 150$ kHz, the channel gain between FL server and the IoV vehicles, $h_{iw} = 2$ dBm, $N_0 =$ -174 dBm/Hz, and the range of transmission power, p_{iw}, is between 1 mW and 10 mW. For the UAVs, similar to [234], we set the transmission power of UAVs, p_{mt}, to be between 0.5 W and 5 W. We also set the bandwidth of the UAVs, $b_m = 400$ kHz, and the channel gain of the FL server and the UAVs, $h_m = 5$ dBm. From Fig. 3.4, we can observe that the communication time needed for the UAVs to transmit the local parameters to the FL server is much lower than that required by the IoV vehicles, hence improving

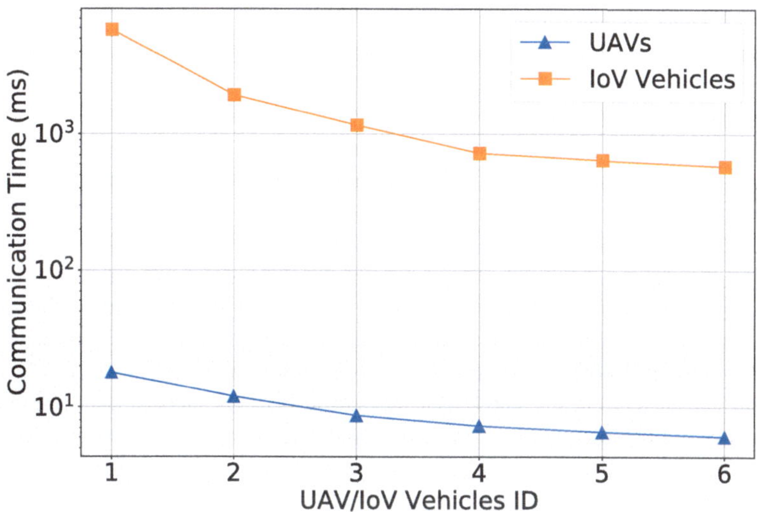

Fig. 3.4 Communication time needed by UAVs and IoV vehicles

Table 3.3 Preference of cells for different coalitions

Cell ID	Coordinates	Importance	Top preference of cells
Cell 1	(200, 300)	18.4	{3}, {1, 3}, {2, 3} or {1, 2, 3}
Cell 2	(500, 700)	25.2	{6}, {2, 6}
Cell 3	(800, 600)	34.6	{6}

the communication efficiency in completing the FL task. Note that even without taking into account of the probability of link failure and the straggler's effect, the communication time required by the IoV vehicles is much larger than that of the UAVs.

3.5.2 *Preference of Cells of Workers*

In order to analyze the preference of cells for the UAVs and to determine their valuations in the bidding process, there are three factors that need to be taken into account, i.e., the importance of the cell, the time required by the UAVs to reach the cell, as well as the ability of the UAVs to complete the FL task. We consider 3 cells where Cell 1, Cell 2, and Cell 3 are located at coordinates $(200, 300)$, $(500, 700)$, and $(800, 600)$, respectively, as presented in the second column of Table 3.3. Column 4 of Table 3.3 shows the top preference of each cell.

Firstly, the cells submit their bids based on their own importance. The more important they are, the higher they earn from the payment by the model owner and thus have higher incentive to employ UAVs to complete the FL task. Based on the sampling rate of the vehicles or RSUs, the importance of the vehicles or RSUs to the FL server is also calculated. Cell 3 is the most important cell with an importance value of 34.6, whereas Cell 1 is the least important cell with an importance value of 18.4. We assume that the price that the model owner pays for each unit of importance is the same for all 3 cells, i.e., $q_1 = q_2 = q_3 = 3$. Secondly, the valuations of each cell also depend on the time required by the UAVs to reach the cell. Cells of workers prefer UAVs that are nearer as they receive higher payment by the model owner for completing the task in a shorter period of time. We assume that the weight parameters for both cell importance and traveling time for the UAVs to the cells to be equal, i.e., $\theta_1 = \theta_2 = 0.5$. The 6 UAVs are randomly distributed, among which their coordinates are presented in Column 2 of Table 3.4. We can calculate the Euclidean distance between cells and UAVs based on the coordinates of them. Since we assume that the UAVs travel at constant velocity v, the time required by UAV m to reach cell i, denoted as t_i^m, can be computed by dividing distance over velocity, i.e., $t_i^m = \frac{d_i^m}{v}$. Thirdly, each cell will only submit bids to the UAVs that can complete the FL task, i.e., fulfill the requirement of completing a certain number of iterations as denoted by the model owner.

To illustrate the preference analysis of the cells of workers, we first consider the scenario in which each UAV facilitates the FL training in the cells individually, i.e., no coalition is formed yet. Table 3.5 illustrates the preference of each cell for each UAV. Cell 1 prefers UAV 3 the most. For Cell 2 and Cell 3, it is clearly seen that both cells prefer UAV 6. This is mainly because UAV 3 is nearest to Cell 1, whereas UAV 6 is nearest to Cell 2 and Cell 3. In addition to that, all cells do not submit any valuation for both UAV 1 and UAV 2 as they cannot individually fulfill the number

Table 3.4 Preference of UAVs for different cells

UAV ID	Coordinates	Energy capacity (Joules)	Preference
UAV 1	(100, 100)	200	–
UAV 2	(300, 500)	500	–
UAV 3	(100, 600)	1000	Cell 3
UAV 4	(600, 200)	1250	Cell 3
UAV 5	(900, 200)	3000	Cell 3
UAV 6	(800, 800)	3500	Cell 3

Table 3.5 Preference of cells for individual UAVs (bolded row implies top preference)

UAV ID	Cell 1	Cell 2	Cell 3
UAV 1	0	0	0
UAV 2	0	0	0
UAV 3	**43.4**	49.9	59.0
UAV 4	39.7	47.6	63.1
UAV 5	34.7	45.6	64.2
UAV 6	34.0	**53.6**	**76.9**

of iterations required for the FL task. This is because UAV 1 and UAV 2 have low energy capacities (as shown in Table 3.4), thereby they are not able to support any cell individually.

However, the allocation of UAVs to the cells solely based on the valuations submitted by the cells may not benefit the UAVs. As an illustration of the single UAV case using Cell 2 as an example, Cell 2 will lose out to Cell 3 in the bidding process for UAV 6 since Cell 3 bids higher for UAV 6. As a result, Cell 2 is left with UAV 3, UAV 4, and UAV 5 to bid for. Among the remaining choices, Cell 2 prefers UAV 3 the most. Thus, Cell 1 will lose out to Cell 2 in bidding for UAV 3 and thus with UAV 4 as the most preferred choice among the remaining UAVs. It may not be economically efficient for UAV 4 to be allocated to Cell 1 as it is far away from Cell 1. The energy needed for UAV 4 to travel to Cell 1 and back to its original position may offset the profit earned in supporting the FL training process in Cell 1. On the other hand, it may be more cost-effective for UAV 1 and UAV 3 to cooperate to support Cell 1 since the energy needed to reach Cell 1 is smaller.

We then consider coalition formation between two or more UAVs. Previously, when only individual UAVs are considered, UAV 1 and UAV 2 are not considered due to the limited energy capacities in completing the training process individually. However, when coalitions can be formed, UAV 1 and UAV 2 are now considered where either of them forms a coalition with UAV 3 or both of them form a coalition with UAV 3. From the perspective of Cell 1, it is indifferent among $\{3\}$, coalitions $\{1, 3\}$, $\{2, 3\}$, and $\{1, 2, 3\}$. The valuations of Cell 1 for these coalitions are the same as the time needed for all the UAVs in the respective coalitions to reach Cell 1 is equal and they are capable of completing the FL task. Note that the time needed for all UAVs in a coalition to reach a cell is determined by the UAV that is farthest from the cell. Similarly, Cell 2 is indifferent between $\{6\}$ and coalition $\{2, 6\}$. Cell 3 prefers UAV 6 since it is capable of completing the task individually and it is nearest to Cell 3.

However, there are two challenges to solve in determining the allocation of UAVs to different cells of workers. Firstly, two cells may prefer the same UAVs. For example, both Cell 2 and Cell 3 prefer UAV 6, but it is not possible for UAV 6 to support both cells at the same time. Secondly, it may not be profitable for some UAVs to support certain cells even if they are preferred by the cells. For example, coalition $\{1,2,3\}$ is equally preferred by Cell 1 as coalitions $\{1, 3\}$, $\{1, 2\}$, and $\{3\}$. However, it may not be profitable for coalition $\{1, 2, 3\}$ to support Cell 1 if a smaller coalition of UAVs can complete the task at a lower cost.

As such, we discuss the profit-maximizing behavior of the UAVs as follows.

3.5.3 *Profit-Maximizing Behavior of UAVs*

To analyze the profit-maximizing behavior of the UAVs, we consider 6 UAVs with different levels of energy capacity, i.e., low, medium, and high (as previously presented in the third column of Table 3.4).

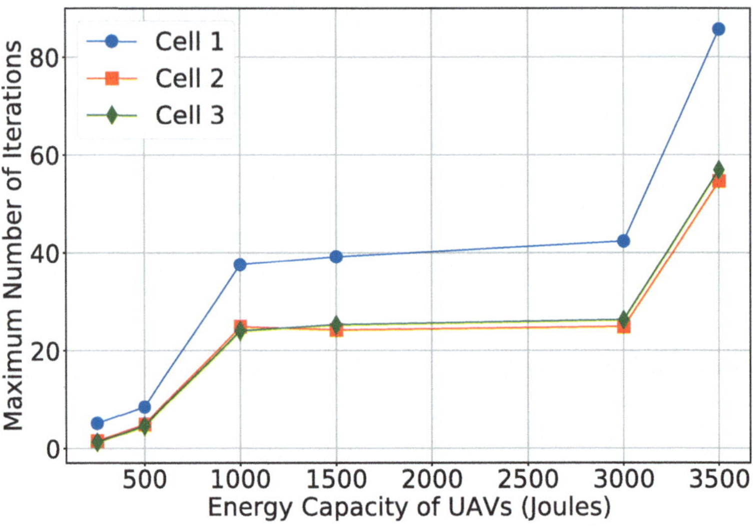

Fig. 3.5 Maximum number of iterations under different energy capacities

Table 3.6 Revenue, cost of profit for different coalitions in each cell

	Cell 1			Cell 2			Cell 3		
Coalition	Rev.	Cost	Prof.	Rev.	Cost	Prof.	Rev.	Cost	Prof.
{3}	43.4	23.5	19.9	49.9	16.2	33.7	59.0	21.5	37.5
{1, 3}	43.4	17.9	25.5	44.7	22.4	22.4	57.7	23.1	34.6
{2, 3}	43.4	32.9	10.5	49.9	32.7	17.2	59.0	31.5	27.5
{1, 2, 3}	43.4	26.8	16.6	44.7	40.0	4.70	57.1	38.4	19.3
{6}	34.0	53.8	−19.8	53.6	33.8	19.8	76.9	34.8	42.1
{2, 6}	34.0	34.3	−0.33	53.6	32.4	21.2	61.7	40.3	21.4

Figure 3.5 shows the maximum number of iterations given the different energy capacities. It is obvious that the larger the energy capacity of the UAV, the larger the number of iterations that the UAV can support. Individually, UAV 1 and UAV 2 do not choose to support any cell as the possible profit earned in supporting any cell is negative. For UAV 1 and UAV 2, since the valuation bids of all cells are zero, it is natural not to support any cell as they are not earning any revenue. Meanwhile, the profit-maximizing property of the UAVs has resulted in the competition of UAV 3, UAV 4, UAV 5, and UAV 6 to facilitate FL training in the same cell, i.e., Cell 3. However, by allocating UAV 3, UAV 4, UAV 5, and UAV 6 to Cell 3, the revenue earned does not increase, but the cost increases which contribute to the overall smaller profit earned. Thus, it is possible to allocate one of the UAVs to another cell such that the overall profit of the UAVs is higher, which is discussed in the next subsection.

Given the varied energy cost of the UAVs and different revenue from different cells, each UAV coalition earns different profits for supporting different cells of workers. From Table 3.6, we observe that although the valuations for coalitions {3}, {1, 3}, {2, 3}, and {1, 2, 3} by Cell 1 are the same, the costs are different. As a result, the profit of coalition {1, 3} is the highest if it is allocated to Cell 1. The addition of UAV 2 to the coalition only incurs extra cost, which leads to a lower profit. Hence, it is not economically viable for coalition {1, 2, 3} to be allocated to Cell 1. Similarly, for Cell 2, coalitions {2, 6} and {6} have the same valuation. It costs lower for Cell 2 to be supported by a coalition of UAV 2 and UAV 6 instead of UAV 6 individually.

With the different revenue and cost structures of the UAVs, we next discuss the auction-based UAV-cell allocation.

3.5.4 Allocation of UAVs to Cells of Workers

Given the valuations of cells for different coalitions of UAVs through the auction, the UAVs as the auctioneers need to decide on the allocation of UAVs to the different cells of workers such that the total profit earned by all UAVs is maximized and the preferences of the cells of workers are taken into account.

The merge-and-split algorithm is used to determine the partition of the UAVs that maximizes the total profit. Figure 3.6 demonstrates the merge-and-split mechanism that determines the partition that increases the total profit earned. When each UAV is

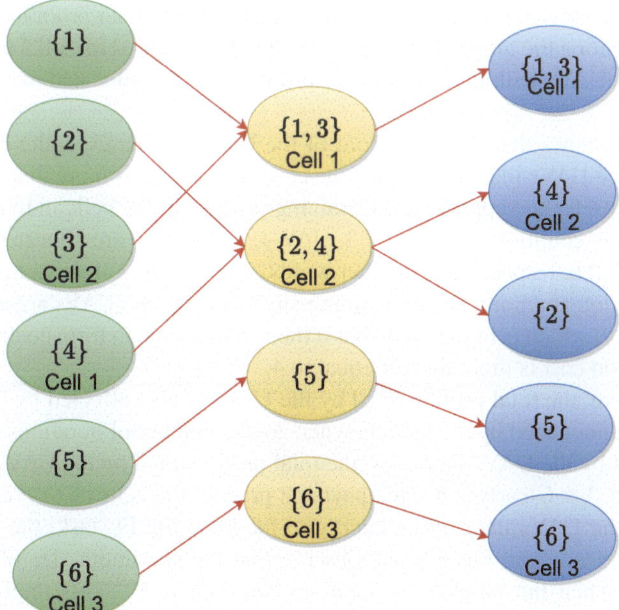

Fig. 3.6 Illustration of merge-and-split mechanism and allocation of UAVs to cells of workers

considered individually where no coalition is formed, the maximum total attainable profit is achieved when UAV 4 is allocated to Cell 1, UAV 3 is allocated to Cell 2, while UAV 6 is allocated to Cell 3. UAV 1, UAV 2, and UAV 5 are not allocated to any cell. The maximum total attainable profit is increased when the partition changes to $\{\{1, 3\}, \{2, 4\}, \{5\}, \{6\}\}$ through a merge mechanism, where UAV 1 and UAV 3 are allocated to Cell 1, UAV 2 and UAV 4 are allocated to Cell 2, and UAV 6 is allocated to Cell 3. Then, the split mechanism results in a change in partition again to $\{\{1, 3\}, \{2\}, \{4\}, \{5\}, \{6\}\}$, where coalition $\{1, 3\}$ is allocated to Cell 1, UAV 4 is allocated to Cell 2, and UAV 6 is allocated to Cell 3. A change in partition will only occur when it results in an increase in total profit earned given the optimal allocation of UAVs to cells.

Although UAV 5 is positively valued by all 3 cells of workers, it is not allocated to any cell. From the perspective of Cell 1, coalition $\{1, 3\}$ can support the FL task at a lower cost than that of UAV 5. Similarly, the costs of UAV 4 and UAV 6, if they are allocated to Cell 2 and Cell 3, respectively, are lower than that if any of the cell is supported by UAV 5. Moreover, UAV 2 is also not allocated to any cell. As mentioned in Sect. 3.5.3, the profit decreases when UAV 2 joins coalition $\{1, 3\}$ in supporting Cell 1, and hence it is more profitable for coalition $\{1, 3\}$ to be allocated to Cell 1, instead of coalition $\{1, 2, 3\}$. UAV 2 also does not join UAV 4 in supporting Cell 2. Although UAV 2 is nearer to Cell 2, it is not capable of completing the FL task individually. The valuations of Cell 2 for UAV 4 and coalition $\{2, 4\}$ are the same. However, it is more cost-effective for Cell 2 to be supported only by UAV 4 instead of both UAV 2 and UAV 4. As such, we see that the grand coalition, where all UAVs join a single coalition, is not stable, thus validating our proof in Sect. 3.3.2. In fact, by forming the grand coalition, it can only be allocated to Cell 3 with a profit of 2.12 as the UAVs will not earn positive profit if they are allocated to either Cell 1 or Cell 2.

Figure 3.7 shows that the total profit of the UAVs decreases as the cooperation cost among the UAVs increases. When the cooperation cost is more than or equal to 4, the UAVs prefer to support the FL training process in the cells individually and not to form any coalition where UAV 3, UAV 4, and UAV 6 are allocated to Cell 2, Cell 1, and Cell 6, respectively. In other words, it is not cost-effective for the UAVs to form coalition anymore. By not forming any coalition, the UAVs are able to earn a profit of 95.7, which is higher than that if the UAVs decide to form coalitions when the cooperation cost is more than or equal to 4.

Furthermore, the total profit earned by the UAVs is also affected by the number of iterations announced by the model owner. As the number of iterations required to be completed by the UAVs increases, the total profit earned by the UAVs decreases as seen in Fig. 3.8. Clearly, since the payment price of the cells of workers does not change and the UAVs incur more energy to facilitate the FL task, the total profit earned decreases. From Fig. 3.8, we observe that the maximum size of coalitions formed is 3 when the number of iterations required is 50. This implies that by forming a coalition of size 3 and be allocated to a cell of workers, the UAV coalition is still able to earn a positive profit. When the number of iterations is 60 to 80, only coalition $\{1, 6\}$ is allocated to Cell 1 where the profit decreases as the number of

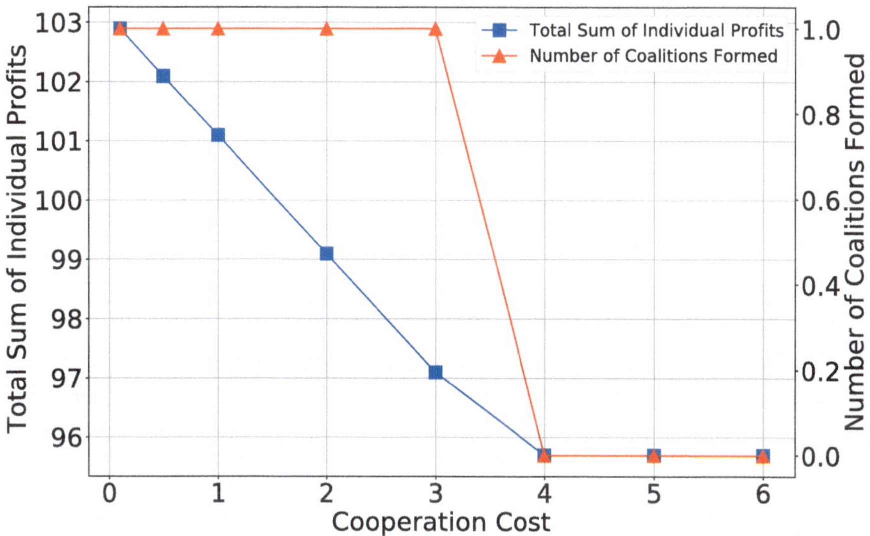

Fig. 3.7 Total profit and number of coalitions vs cooperation cost

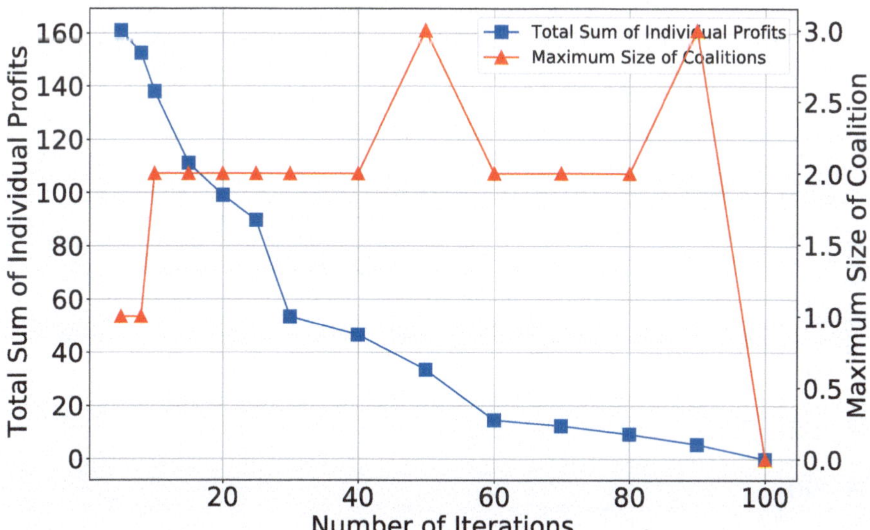

Fig. 3.8 Total profit and size of coalitions vs number of iterations

iterations increases. The rest of the UAVs are allocated to neither Cell 2 nor Cell 3 as the profits earned by any possible coalitions from supporting any of these cells are negative. When the number of iterations required by the FL task is 90, the coalition {1, 3, 6} is allocated to Cell 1. The UAVs prefer not to support any cell of workers, and thus there is no need to form a coalition when the number of iterations is 100.

Therefore, a coalition with size of larger than 3 does not form due to two reasons. Firstly, it is possible to form a smaller coalition to support a cell of workers such that there is no need for a larger UAV coalition. Secondly, the cost incurred by a larger UAV coalition is larger than the revenue gained such that the UAVs are better off not supporting any cell of workers.

Next, we compare the performance of the proposed joint auction–coalition formation framework against the existing schemes.

3.5.5 Comparison with Existing Schemes

For comparison, we have chosen two existing schemes: (1) the merge-and-split algorithm to determine the formations of the UAV coalitions with random allocation to the IoV groups and (2) random partitioning of the UAV coalitions with second-price auction. We have run three rounds of simulations and the performances of the three schemes are shown in Fig. 3.9.

The joint auction–coalition formation framework achieves the highest profit in all three rounds of simulations. Due to the randomness in either the partitioning or the allocation of the UAV coalitions, the UAVs are not able to maximize their profits. On the one hand, by adopting the merge-and-split framework to determine the formations of the UAV coalitions and randomly allocating the UAV coalitions to the IoV groups, the profits of the UAV coalitions are not maximized as the UAV

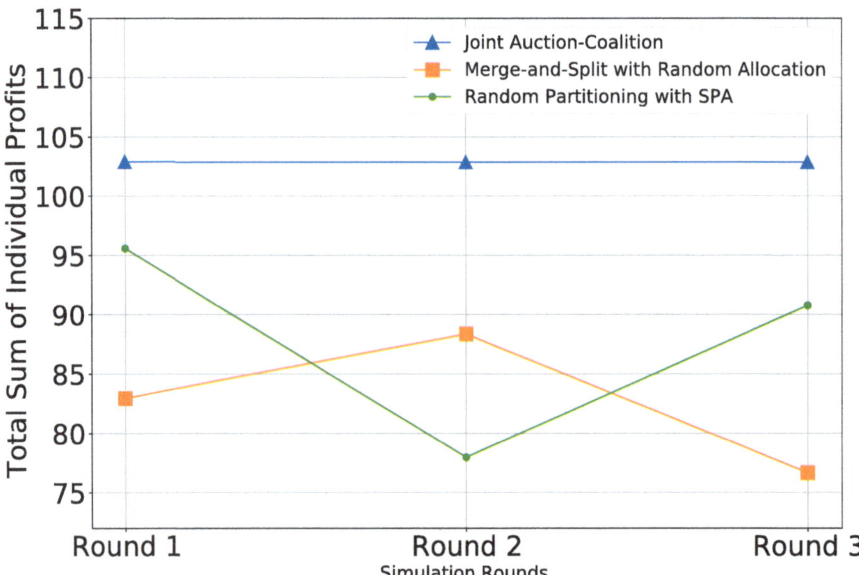

Fig. 3.9 Comparison with existing schemes

coalitions are not allocated to the IoV groups that value them most. On the other hand, by randomly deciding on the partitioning of the UAV coalitions and allocating them based on the second-price auction, the potential of the UAV coalitions to earn higher profits is not exploited, which is achievable by considering other forms of partitioning.

As the number of IoV components increases, the allocated UAV coalition incurs a higher communication cost. However, the algorithm complexity of the proposed joint auction–coalition formation framework is not affected as it does not depend on the number of workers. As such, even when the vehicles are constantly moving and thus the cell formations vary over time, the proposed algorithm remains efficient as long as the number of cells formed remains small. As discussed in Sect. 3.4.4, the overall time complexity of the proposed algorithm depends on both the number of merge-and-split attempts and the complexity of the auction-based allocation. Hence, the complexity of the proposed algorithm increases significantly when the number of UAVs and formed cells in the network increases. In the future, ways to increase the algorithm efficiency for a large number of UAVs and cells can be explored.

3.6 Conclusion and Chapter Discussion

In this chapter, we have proposed a joint auction–coalition formation framework of UAVs to facilitate resource-constrained IoV components in completing the FL tasks. Firstly, we design an auction scheme where each cell of workers submit its bids to all possible coalitions of UAVs based on their importance of the cells and the distance between the cells and the UAVs. Then, we use the merge-and-split algorithm to decide on the optimal coalitional structure that maximizes the total profit of the UAVs.

For future works, we can consider more factors in determining the coalition formation of UAVs. As discussed in Sect. 1.5.2, security issues may impede the successful implementation of FL. With additional workers involved in this chapter (i.e., UAVs), it is important we consider that not some UAVs may be malicious. As such, we can take into account the social properties of the UAVs when they form coalitions such that malicious UAVs can be identified and eliminated so that the FL performance is not adversely affected.

Secondly, this chapter has considered simplistic aggregation mechanisms for the ease of discussion. For further improvement of the communication efficiency of FL, we may consider communication and aggregation schemes such as over-the-air aggregation discussed in Sect. 1.4.2.

Thirdly, we have utilized the SPA in this chapter. However, as will be discussed further in Chap. 4, the SPA has its own drawbacks. We will further introduce the readers to alternative approaches for service pricing.

Chapter 4
Evolutionary Edge Association and Auction in Hierarchical Federated Learning

4.1 Introduction

Today, the predominant approach for AI based model training is cloud-centric, i.e., the data owners transmit the training data to a public cloud server for processing. However, this is no longer desirable due to the following reasons. Firstly, privacy laws, e.g., GDPR [235], are increasingly stringent. In addition, the privacy-sensitive data owners can opt out of data sharing with third parties. Secondly, the transfer of massive quantities of data to the distant cloud burdens the communication networks and incurs unacceptable latency especially for time-sensitive tasks. As such, this necessitates the proposal of Edge Computing [17] as an alternative, in which raw data are processed at the edge of the network, closer to where data are produced.

The confluence of Edge Computing and AI gives rise to Edge Intelligence, which leverages the storage, communication, and computation capabilities of end devices and edge servers to enable edge caching, model training, and inference [236] closer to where data are produced. One of the enabling technologies [237] of Edge Intelligence is the privacy-preserving machine learning paradigm termed FL [21]. In FL, only the updated model parameters, rather than the raw data, need to be transmitted back to the model owner for global aggregation. The main advantages of FL are (a) FL enables privacy-preserving collaborative machine learning, (b) FL leverages on the computation capabilities of IoT devices for local model training, thus reducing the computation workload of the cloud, and (c) model parameters are often smaller in size than raw data, thus alleviating the burden on backbone communication networks. This has enabled several practical applications.

However, the FL network is envisioned to involve thousands of heterogeneous distributed devices, e.g., smartphones and Internet of Thing (IoT) devices [108]. In this case, communication inefficiency remains a key bottleneck in FL. Specifically, node failures and device dropouts due to communication failures can lead to inefficient FL. Moreover, workers, i.e., data owners, with severely limited connectivity, are unable to participate in the FL training, thus adversely affecting the model's

W. Y. B. Lim et al., *Federated Learning Over Wireless Edge Networks*, Wireless Networks, https://doi.org/10.1007/978-3-031-07838-5_4

ability to generalize. As such, solutions from edge computing have recently been incorporated to solve the communication bottleneck in FL. In [237–239], a *hierarchical* FL (HFL) framework is proposed in which the workers do not communicate directly with a central controller, i.e., the model owner. Instead, the local parameter values are first uploaded to edge servers, e.g., at base stations, for intermediate aggregation. Then, communication with the model owner is further established for global aggregation. Besides reducing the instances of global communications with the remote servers of the model owner, this relay approach reduces the dropout rate of devices.

While [238] discusses convergence guarantees and presents empirical results to show that the HFL approach does not compromise on model performance, the challenges of resource allocation and incentive mechanism design have not yet been well addressed in the HFL framework. In 5G and Beyond networks, the resource sharing and incentive mechanism design for end–edge–cloud collaboration are of paramount importance to facilitate efficient Edge Intelligence [237].

In this chapter, we consider a decentralized learning based system model inspired by the HFL. In our system model, there exist data owners, hereinafter referred to as workers, that participate in the FL model training facilitated by different cluster heads, e.g., base stations that support the intermediate aggregation of model parameters and efficient relaying to the model owners (Fig. 4.1). We consider a two-level resource allocation and incentive design problem as follows:

1. *Lower level (between workers and cluster heads):* Each worker can freely choose which cluster to join. To encourage the participation of workers, the cluster heads

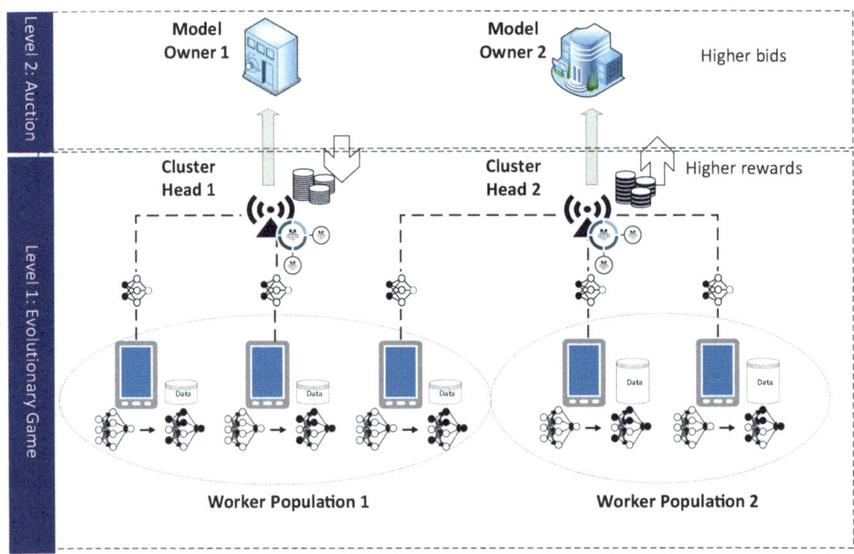

Fig. 4.1 An illustration of the hierarchical system model

offer reward pools to be shared among workers based on their data contribution in the cluster. For example, a worker that has contributed more data[1] during its local training will receive a larger share of the reward pool. Moreover, the cluster heads offer the workers' resource blocks, i.e., bandwidth, to facilitate the efficient uplink transmission of the updated model parameters. However, as more workers join a cluster, the payoffs are inevitably reduced due to the division of rewards over a larger number of workers and the increased communication congestion. Thus, the cluster selection strategies of each worker can affect the payoffs of other workers. Accordingly, the workers may slowly adapt their strategies in response to other workers. In contrast to conventional optimization approaches, we use the evolutionary game theory [240] to derive the equilibrium composition of the clusters. Our game formulation enables bounded rationality and worker dynamics to be captured. Specifically, the workers gradually adapt their strategies in response to other non-cooperative workers. To achieve their objectives, they observe each other's strategies and gradually adjust their strategies accordingly. The solution is therefore not immediately derived.

2. *Upper level (between cluster heads and model owners):* There may be multiple model owners in the network that aim to train a model for their respective usage collaboratively with the participation of the workers and cluster heads. However, at any point in time, each worker and cluster head can only participate in the training process with a single model owner. To derive the allocation of cluster head to the model owner, as well as the optimal pricing of the services of the cluster head by the competitive model owners, we adopt a deep learning based auction mechanism that preserves the properties of truthfulness of the bidders, while simultaneously achieving revenue maximization for the cluster heads.

The main contributions of this chapter are as follows:

1. We propose a joint resource allocation and incentive design framework for the HFL. The "Edge for AI" [241] approach supports decentralized Edge Intelligence, i.e., FL at the edge with reduced reliance on a central controller.
2. We model the cluster selection decisions of the workers as an evolutionary game. Then, we provide proofs for the uniqueness and stability of the evolutionary equilibrium. In contrast to conventional optimization tools that assume that the players are perfectly rational, our model enables us to capture the dynamics and bounded rationality of player decisions.
3. To assign the cluster heads to model owners, we use a deep learning based auction mechanism. In contrast to conventional auctions, the deep learning based auction ensures seller revenue maximization while satisfying the individual rationality and incentive compatibility constraints.

[1] Note that this knowledge is available to the cluster heads given that the workers have to report their available resources during the client selection procedure [88]. In this chapter, we omit discussions regarding the client selection phase.

The organization of this chapter is as follows. In Sect. 4.2, we discuss the system model and problem formulation. In Sect. 4.3, we study and analyze the evolutionary game. In Sect. 4.4, we discuss the deep learning based auction mechanism. Section 4.5 provides the performance evaluation and Sect. 4.6 concludes the chapter.

4.2 System Model and Problem Formulation

4.2.1 System Model

We consider a network that consists of a set $\mathcal{N} = \{1, \ldots, n, \ldots, N\}$ of workers. There exist L distinct model owners, each of which desires to develop an AI model for its own purposes, e.g., traffic crowdsensing [183] or location based recommendation [242]. Given the communication constraints of individual workers, the central controller-reliant conventional FL architecture is prone to high device dropout rates [239]. Moreover, we have J cluster heads, e.g., base stations, employed across the network to facilitate the HFL task, the set of which is denoted by $\mathcal{J} = \{1, \ldots, j, \ldots, J\}$. Each worker can choose to associate with anyone $j \in \mathcal{J}$ cluster head.

Without loss of generality, each cluster head $j \in \mathcal{J}$ can only serve a single model owner $l \in \mathcal{L}$ at a time and facilitates the HFL process for a cluster j of $p_j N$ workers, where $\sum_j^J p_j = 1$.

In HFL, a cluster head j first receives an initial global model, i.e., parameters denoted by the vector $\boldsymbol{w}$, from a model owner that has chosen its services. It then relays the global model to its workers. The FL model training takes place over K iterations to minimize the global loss $F^K(\boldsymbol{w})$, where K is stipulated by the model owner. Each kth iteration, $k \in \{1, \ldots, k, \ldots, K\}$, consists of three steps [65] namely:

1. *Local Computation*: Each worker trains the received global model $\boldsymbol{w}^{(k)}$ locally.
2. *Wireless Transmission*: The worker transmits the model parameter update to its cluster head j.
3. *Intermediate Model Parameter Update*: All parameter updates received from its $p_j N$ workers are aggregated by the cluster head to derive an updated intermediate model $\boldsymbol{w}_j^{(k+1)}$, which is then transmitted back to the worker for the $(k+1)$th training iteration.

After K iterations, the intermediate model $\boldsymbol{w}_j^{(K)}$ is transmitted to the model owner for global aggregation with the intermediate parameters collected from other clusters. A new set of updated global parameters is derived by the model owner that sends it out to its cluster heads for another round of local model training.

In this chapter, we assume that the cluster heads are predetermined, e.g., through the cluster head selection algorithms based on energy efficiency [243, 244], trust [245], and social effects [246]. Instead, we focus our study on a two-level optimization problem as follows: (a) in the *lower level*, we adopt an evolutionary game approach to study the dynamics of cluster selection by the workers to derive the dynamics of the composition of each cluster and (b) in the *upper level*, we adopt a deep learning based auction to value each cluster head's worth to a model owner.

4.2.2 Lower-Level Evolutionary Game

In the lower level, the cluster formation is derived given J cluster heads. Each cluster head has the objective of attracting more workers to join its cluster, since this ensures that the cluster will have a larger *data coverage* across the network. With a larger data coverage, the cluster value is increased, e.g., due to the fact that the model performance increases with more training data [235].

To encourage the participation of the workers, each cluster head offers a reward pool to be shared by all workers in the cluster. The reward to be distributed to each worker is based on the proportion of the worker's contributions in the cluster, i.e., its data quantity relative to the total amount of data in the cluster. On the one hand, a cluster that offers a high reward pool is more attractive to workers. On the other hand, when more workers join that cluster, the reward pool has to be shared among a larger number of workers. Thus, the worker decisions as to which cluster they choose to join are interrelated with the decisions of other workers. We adopt an evolutionary game theory approach to model the dynamics of cluster formation.

4.2.3 Upper-Level Deep Learning Based Auction

The lower-level evolutionary game gives us the data coverage of each cluster. For example, a cluster head that has greater data coverage will be deemed more valuable to an FL model owner, since the model performance, e.g., inference accuracy, is improved [113]. However, recall from Sect. 4.2.1 that there exists more than one model owner in the network seeking to secure the services of the cluster head to facilitate the HFL. In consideration of the competition among model owners, we adopt an auction mechanism in which L model owners bid for the services of each cluster head $j \in \mathcal{J}$. Specifically, we utilize the deep learning based auction mechanism which has the attractive properties of ensuring the truthfulness of the bidders, as well as revenue maximization for the seller (i.e., cluster head), as discussed in Sect. 4.4.

4.3 Lower-Level Evolutionary Game

4.3.1 Evolutionary Game Formulation

In the following, we formulate the cluster selection as an evolutionary game:

1. *Players:* The set of workers $\mathcal{N} = \{1, \ldots, n, \ldots, N\}$ in the FL network are the players of the evolutionary game. For clarity, we use the term "workers" hereinafter.
2. *Population:* We partition the workers into $\mathcal{M} = \{1, \ldots, m, \ldots, M\}$ populations based on the data quantities[2] that each worker owns or the data coverage proportion of the worker using conventional data mining tools, e.g., k-means. The data coverage proportion can be a reflection of the workers' market share in the case when organizations are considered, e.g., based on the proportion of users a bank has, or usage frequency in the case in which individuals are considered, e.g., based on how often the worker uses an IoT device. Each worker of population m owns d_m training data samples, whereas the total data quantity in a population is denoted D_m. We denote the number of workers in each population as $n_m = p_m N$ where $p_m \in [0, 1]$ and $\sum_{m=1}^{M} p_m = 1$. In other words, we have M populations of workers in the network, where all workers within a population own the same number of data samples.
3. *Strategy:* The strategy of each worker in population m is the selection of a cluster to join so as to achieve utility maximization. The strategy space of each worker n in population m is denoted by $\mathcal{S}_n^{(m)} = \{a_{n,1}^{(m)}, \ldots, a_{n,j}^{(m)}, \ldots, a_{n,J}^{(m)}\}$ in which $a_{n,j}^{(m)}$ is a binary variable where $a_{n,j}^{(m)} = 1$ represents that the worker n in population m chooses the cluster j, whereas $a_{n,j}^{(m)} = 0$ indicates otherwise.
4. *Population Share:* We denote the fraction of population m that selects strategy j, i.e., cluster j, by $x_j^{(m)}$ where $\sum_{j=1}^{J} x_j^{(m)} = 1$. The population state [247] is denoted by the vector $\mathbf{x}^{(m)} = [x_1^{(m)}, \ldots, x_j^{(m)}, \ldots, x_J^{(m)}]^T \in \mathbb{X}$.
5. *Payoff:* The expected payoff of each worker is determined by its net utility, which is the difference between the reward that it derives from joining a cluster, and the cost of participating in the FL model training. We further discuss payoffs in Sect. 4.3.2.

As an illustration, the system model and game formulation are illustrated in Fig. 4.1, for the case of two populations. Each worker in population 1 has fewer data samples than each worker in population 2. Each worker in the population can also choose to join either cluster head. Eventually, each cluster head is associated

[2] For simplicity, we only consider data quantity as the criterion for clustering here. Note that we may cluster the workers based on other important factors such as geolocation (as a proxy for data distribution) and reputation metrics (as a proxy for quality of contribution). Our work can easily be extended to cover such measures.

with a certain level of data coverage and has its worth evaluated using the auction mechanism discussed in Sect. 4.4.

Note that in this chapter, we consider that each worker can only join a single cluster, as the worker device is unable to support two instances of model training in parallel. However, our model can be extended to the situation in which each worker can join more than one cluster at a time. In this case, the worker can be modeled to decide, in an evolutionary process, on how its limited resources can be divided among the model owners. Then, $x_j^{(m)} \in [0, 1]$ is denoted to represent the share of resources a worker from population m contributes to cluster head j, where $\sum_{j=1}^{J} x_j^{(m)} = 1$.

4.3.2 *Worker Utility and Replicator Dynamics*

The rewards derived by workers of population m, from joining a cluster j for K iterations of FL model training, are given by

$$p_j^{(m)} = \alpha_j \frac{x_j^{(m)} D_m}{\sum_{m=1}^{M} x_j^{(m)} D_m} + R_j, \tag{4.1}$$

where α_j is the reward pool to be divided across all workers in cluster j based on their data contributions, $\frac{x_j^{(m)} D_m}{\sum_{m=1}^{M} x_j^{(m)} D_m}$ is the share of rewards based on the worker's data contribution,[3] and R_j is a fixed reward offered to workers in cluster j based on the compensation for the workers' participation costs.

The cost of workers of population m incurred from joining a cluster j is given by the addition of the computation and communication cost[4] over the K iterations of the model training. The computation cost is as follows:

$$c_m^{cmp} = \eta \kappa \theta_m f_m^2, \tag{4.2}$$

where η is the unit cost of energy consumption, κ is the coefficient of the value that is determined by the circuit architecture of the worker Central Processing Unit (CPU) [171], θ_m is the number of CPU cycles required to perform local computation, i.e., model training, and f_m^2 refers to the computation capability of

[3] In this chapter, we make a simplifying assumption that the workers are not malicious. In practice, the parameter contributions of the workers can be randomly verified, e.g., detecting model parameters with outlying magnitudes, and malicious workers with anomalous contributions can be penalized from participating in the future FL training.

[4] For ease of presentation, we have used a simple computation and communication cost model in this chapter. However, a more specific modeling of the cost can be referenced from Chap. 2. The use of the model does not affect the findings of this chapter.

the worker which is determined by the clock frequency of the worker CPU. Without loss of generality, we have the computation cost held constant throughout for all workers, i.e., $c_m^{cmp} = c^{cmp}, \forall m \in \mathcal{M}$. To account for the varying computation and communication capabilities, we can straightforwardly extend our work to include multiple heterogeneous populations with varying computation capabilities. For example, if there are Λ varying computation costs, we can have ΛM populations accordingly.

The main benefit of HFL is that devices with communication constraints are able to participate in FL. To facilitate the communication of parameters, the cluster head, e.g., a base station, distributes communication resource blocks to all participants within the cluster. On the one hand, clusters with more communication resources are attractive to participants since the participants can benefit from a higher achievable uplink transmission rate. On the other hand, with more participants attracted to join the cluster, the increased competition for resource blocks leads to more congestion. As such, following [248], we model the disutilities arising from network congestion effects as

$$c_{m,j}^{com}(t) = \zeta_j \left(\sum_{m \in \mathcal{M}} x_j^{(m)}(t) \right)^2, \tag{4.3}$$

where ζ_j is the congestion coefficient determined by the resource constraints of the cluster head [248], whereas $\left(\sum_{m \in \mathcal{M}} x_j^{(m)}\right)^2$ represents the usage profile across populations in the network for a particular cluster. Specifically, a cluster head with more resources will have a lower congestion coefficient. Moreover, workers in a less-populated cluster experience less congestion.

The total cost of participation incurred by worker n_m (of population m) in cluster j is obtained as

$$c_j^{(m)}(t) = c^{cmp} + c_{m,j}^{com}(t). \tag{4.4}$$

At time t, the net utility that the workers in class m receive for their participation in cluster j is

$$u_j^{(m)}(t) = \mathcal{U}\left(p_j^{(m)}(t) - c_j^{(m)}(t)\right), \tag{4.5}$$

where we assume $\mathcal{U}(\cdot)$ to be a linear utility function indicating the risk neutrality of workers without loss of generality [249, 250].

Accordingly, at time t, the average utility of workers in population m across all J clusters is

$$\bar{u}^{(m)}(t) = \sum_{j=1}^{J} x_j^{(m)} u_j^{(m)}(t). \tag{4.6}$$

In practice, information regarding the utility derived from joining different clusters can be exchanged and compared among workers within the network [251]. Workers may thus switch from one cluster to another to seek higher utilities. To capture the dynamics of the cluster selection and model the strategy adaptation process, we define the replicator dynamics [252] as follows:

$$\dot{x}_j^{(m)}(t) = f_j^{(m)}(\mathbf{x}^{(m)}(t)) = \delta x_j^{(m)}(t)\left(u_j^{(m)}(t) - \bar{u}^{(m)}(t)\right), \quad \forall m \in \mathcal{M}, \forall j \in \mathcal{J}, \forall t, \tag{4.7}$$

where δ refers to the positive learning rate of the population that controls the speed at which workers adapt their strategies. For example, in a network with communication bottlenecks [32] or negative network effects [26], the learning rate tends to be slower as the worker requires more time to collect the information required to change its decision.

The replicator dynamics is a series of ordinary differential equations (ODEs) with the initial condition $\mathbf{x}^{(m)}(0) \in \mathbb{X}$ [253]. Specifically, based on the replicator dynamics, workers in population m can adapt their strategy, i.e., switch from one cluster to another if their utilities are lower than the expected utility. The *evolutionary equilibrium* is a fixed point in (4.7) that is reached in a particular t when $\dot{x}_j^{(m)}(t) = 0, \forall m \in \mathcal{M}, \forall j \in \mathcal{J}$. In other words, at the evolutionary equilibrium, workers from all clusters derive an identical payoff such that there is no longer a need to deviate from their prevailing clusters.

In a dynamic system, it is of paramount importance that the equilibrium is stable and unique. In terms of stability, an evolutionary equilibrium remains to be $\dot{x}_j^{(m)}(t) = 0$ for all time periods after the equilibrium is first reached. In terms of uniqueness, the same evolutionary equilibrium is reached regardless of the initial conditions. In Sect. 4.3.3, we prove the existence, uniqueness, and stability of the solution to (4.7).

4.3.3 Existence, Uniqueness, and Stability of the Evolutionary Equilibrium

In this section, we first prove the boundedness of (4.7) in Lemma 4.1.

Lemma 4.1 *The first-order derivatives of $f_j^{(m)}(\mathbf{x}^{(m)}(t))$ with respect to $x_v^{(m)}(t)$ is bounded for all $v \in J$.*

Proof For ease of presentation, we omit the notations of t and (m) in this proof. The first-order derivative of $f_j(\mathbf{x})$ with respect to x_v, where $v \in \mathcal{J}$, is given by

$$\frac{\mathrm{d}f_j(\mathbf{x})}{\mathrm{d}x_v} = \delta\left[\frac{\mathrm{d}x_j}{\mathrm{d}x_v}\left(u_j - \bar{u}\right) + x_j\left(\frac{\mathrm{d}u_j}{\mathrm{d}x_v} - \frac{\mathrm{d}\bar{u}}{\mathrm{d}x_v}\right)\right]. \tag{4.8}$$

For ease of notation, denote $A(\mathbf{x}_j) = \sum_{m=1}^{M} x_j D_m$. Next, we derive $\frac{\mathrm{d}u_j}{\mathrm{d}x_v}$ as follows:

$$\frac{\mathrm{d}u_j}{\mathrm{d}x_v} = \alpha_j \left(\frac{\frac{\mathrm{d}x_j}{\mathrm{d}x_v} D_m}{A(\mathbf{x}_j)} - \frac{x_j D_m^2}{A^2(\mathbf{x}_j)} \right) - 2\zeta_j \left(\sum_{m \in \mathcal{M}} x_j \right). \tag{4.9}$$

It follows that $\left|\frac{\mathrm{d}u_j}{\mathrm{d}x_v}\right|$ and thus $\left|\frac{\mathrm{d}\bar{u}}{\mathrm{d}x_v}\right|$ are clearly bounded $\forall v \in \mathcal{J}$. Therefore, this represents that $\left|\frac{\mathrm{d}f_j(\mathbf{x})}{\mathrm{d}x_v}\right|$ is bounded. This proof also applies to all M populations and T time periods. □

Theorem 4.1 *For any initial condition $\mathbf{x}^{(m)}(0) \in \mathbb{X}$, there exists a unique evolutionary equilibrium to the dynamics defined in (4.7).*

Proof From Lemma 1, we have proven that the replicator dynamics $f_j^{(m)}(\mathbf{x}^{(m)}(t))$ is bounded and continuously differentiable $\forall \mathbf{x}^{(m)}(t) \in \mathbb{X}, \forall m \in \mathcal{M}, \forall j \in \mathcal{J}, \forall t$. Therefore, the maximum absolute value of its partial derivative given in (8) is a Lipschitz constant. According to the Mean Value Theorem, there exists a constant c between $x_1^{(m)}(t)$ and $x_2^{(m)}(t)$ such that $\frac{\left|f_j^{(m)}(x_1^{(m)}(t)) - f_j^{(m)}(x_2^{(m)}(t))\right|}{(x_1^{(m)}(t) - x_2^{(m)}(t))} = \frac{\mathrm{d}f_j(c)}{\mathrm{d}x_v}$. Therefore, we can define the relation

$$\begin{aligned} \left|f_j^{(m)}(x_1^{(m)}(t)) - f_j^{(m)}(x_2^{(m)}(t))\right| &\le \Gamma \left|x_1^{(m)}(t) - x_2^{(m)}(t)\right|, \\ &\forall (x_1^{(m)}, x_2^{(m)}) \in \mathbb{X}, \forall m \in \mathcal{M}, \forall t, \end{aligned} \tag{4.10}$$

where $\Gamma = \max\left\{\left|\frac{\mathrm{d}f_j(c)}{\mathrm{d}x_v}\right|\right\}$. Following the Lipschitz condition [254], this implies that the replicator dynamics, i.e., an initial value problem with $\mathbf{x}^{(m)}(0) \in \mathbb{X}$, in (7) has a unique solution $x_j^{(m)\star} \in \mathbb{X}$. □

Next, we prove the stability of the evolutionary equilibrium in the following theorem.

Theorem 4.2 *For any initial condition $\mathbf{x}^{(m)}(0) \in \mathbb{X}$, the evolutionary equilibrium to the dynamics defined in (4.7) is stable.*

Proof We define the Lyapunov function

$$G(\mathbf{x}^{(m)}(t)) = \left(\sum_{m=1}^{M} \sum_{j=1}^{J} x_j^{(m)}(t) \right)^2, \tag{4.11}$$

which is positive definite since

$$G(\mathbf{x}^{(m)}(t)) \begin{cases} = 0 & \text{if } \mathbf{x}(t) = \mathbf{0} \\ > 0 & \text{otherwise.} \end{cases} \tag{4.12}$$

Taking the first-order derivative with of $G(\mathbf{x}^{(m)}(t))$ with respect to t,

$$\frac{\mathrm{d}G(\mathbf{x}^{(m)}(t))}{\mathrm{d}t} = 2\left(\sum_{m=1}^{M}\sum_{j=1}^{J} x_j^{(m)}(t)\right)\left(\sum_{m=1}^{M}\sum_{j=1}^{J} \dot{x}_j^{(m)}(t)\right). \tag{4.13}$$

Note that at any point of time, $\sum_{m=1}^{M}\sum_{j=1}^{J} x_j^{(m)}(t) = M$. Thus, the replicator dynamics have to equate to zero for this to hold, i.e., the net movements and strategy adaptations across clusters are zeroed in order for the population to remain constant. Specifically,

$$\sum_{m=1}^{M}\sum_{j=1}^{J} \dot{x}_j^{(m)}(t) = 0, \forall t. \tag{4.14}$$

Therefore, (4.14) ensures that $\frac{\mathrm{d}G(\mathbf{x}^{(m)}(t))}{\mathrm{d}t} = 0$, which satisfies the Lyapunov conditions required for stability, as defined in Lyapunov's second method for stability [255]. □

As such, we have proven the uniqueness and stability of the evolutionary equilibrium.

Next, we discuss the procedures to derive the equilibrium cluster data coverage based on the replicator dynamics in (4.7). In contrast to the population evolution algorithm [251] which involves the intervention of a centralized controller, e.g., in disseminating information of potential payoffs that can be derived from joining a particular cluster, we consider the implementation of a decentralized cluster selection algorithm in Algorithm 4.1.

At the initialization phase, workers in each population m are randomly assigned to j clusters, where $m \in \mathcal{M}$, $j \in \mathcal{J}$. At each time period t, the workers compute their utilities and the average utility of workers in the population. Note that in practice, the workers may not have the complete information of all workers belonging to the same population in a large network. As such, its knowledge of the average utility in the population is based on the worker's "best guess," i.e., the expected average utility. This procedure is simply a comparison between (a) the worker's own utility from joining a particular cluster j, i.e., $u_j^{(m)}(t)$ and (b) the expected average utilities of other workers from the same population which has chosen to join other clusters, i.e., $E(\bar{u}^{(m)}(t))$. Thereafter, the evolution of the population state can be derived following the replicator dynamics.

The output of Algorithm 4.1 is the population state that is observed after t_{max} iterations. Then, we are eventually able to derive the data coverages of the cluster

Algorithm 4.1 Cluster selection for HFL

Input: Worker and cluster characteristics
Initialization: Workers in population m each assigned to a random cluster
while $t < t_{max}$ **do**
    for $m \in \mathcal{M}$ **do**
        Payoff Computation
        for $j \in \mathcal{J}$ **do**
            Derive $u_j^{(m)}(t) = p_j^{(m)}(t) - c_j^{(m)}(t)$
        end for
        Compute $E(\bar{u}^{(m)}(t)) = E(\sum_{j=1}^{J} x_j^{(m)} u_j^{(m)}(t))$
        Cluster Selection
        for $j \in \mathcal{J}$ **do**
            Derive $\dot{x}_j^{(m)}(t) = \delta x_j^{(m)}(t)\left(u_j^{(m)}(t) - \bar{u}^{(m)}(t)\right)$ and $x_j^{(m)}(t)$
        end for
    end for
end while
for $j \in \mathcal{J}$ **do**
    Compute $D_j = \sum_{m=1}^{M} x_j^{(m)}(t_{max}) D_m$
end for

head, i.e., the proportion of data across the network that each cluster head can cover, as follows:

$$D_j = \sum_{m=1}^{M} x_j^{(m)}(t_{max}) D_m. \tag{4.15}$$

4.4 Deep Learning Based Auction for Valuation of Cluster Head

4.4.1 Auction Formulation

Based on the cluster formations from the evolutionary game, we are able to derive the data coverage D_j of the cluster head $j \in \mathcal{J}$.

As each cluster head can only offer its services to a single model owner, i.e., the workers' participation in the FL model training, the model owners need to compete for the services of the cluster heads. Each model owner $l \in \mathcal{L}$ has a different preference for the accuracy of their models, e.g., applications for accident warning and prediction [185] require higher accuracy than the route planning and navigation systems. Following the work in [256], the FL model accuracy A_l of model owner l can be expressed as a power law function that is denoted as follows:

$$A_l(\mu_l) = \sigma - \upsilon \mu_l^{-r}, \tag{4.16}$$

where σ, v, and r are parameters that can be calibrated depending on the model to be trained. μ_l is the data coverage required by model owner l to achieve its required model accuracy, σ is the upper bound of the accuracy that can be derived from historical data, whereas v and r are the fixed parameters of the function.

In general, when the requirement for data coverage of the model owner μ_l is larger, the model owner has more incentive to pay a higher price for the services of the cluster heads which have more data coverage. In contrast, a model owner that already has some preexisting training data d_l will have less incentive to bid for the services of a cluster head. Therefore, the valuation b_l of model owner l for the services offered by the cluster heads can be expressed as $b_l = \mu_l - d_l$.

In order to maximize the revenue of the cluster heads and to ensure that the services from the cluster heads are allocated to the model owners that value them most, we model the allocation problem as multiple rounds of single-item auctions. In this auction, the cluster heads are the sellers, i.e., auctioneers, while the model owners are the buyers, i.e., bidders. The cluster head with the highest amount of data coverage is the first to auction its services to the model owners. All model owners submit their bids to compete for the service. Then, the cluster head collects the bid profile $(b_1, \ldots, b_l, \ldots, b_L)$ to decide on the winning model owner l^* and the corresponding payment price $\theta_{l^\star}$. After each round of auction, the winning model owner has higher data coverage, i.e., higher d_l. Thus, its valuation in the next round of auction naturally has to be updated and decreases. The auction ends when all cluster heads have been allocated to the model owners. Note that the model owners may participate in more than one round of auction to fulfill their data coverage requirement, e.g., if the data coverage that a single cluster has is insufficient to fulfill its needs.

Accordingly, the utility of the model owner in each round of the auction is as follows:

$$u_l = \begin{cases} b_l - \theta_{l^\star} & \text{if the model owner wins the bid,} \\ 0 & \text{otherwise.} \end{cases} \tag{4.17}$$

An optimal auction [257] has two characteristics:

1. *Individual Rationality (IR):* By participating in the auction, the model owners receive non-negative payoff, i.e., $u_l \geq 0$.
2. *Incentive Compatibility (IC):* There is no incentive for the model owners to submit bids other than their true valuations, i.e., the bidders always bid truthfully.

In each round of the auction, in order to determine the payment price $\theta_{l^\star}$ of the winning model owner, traditional auction schemes such as the first-price auction and second-price auction (SPA) may be adopted. However, each of these traditional auction schemes has its own drawbacks.

The traditional first-price auction, in which the highest bidder pays the exact bid it submits, maximizes the revenue of the seller but does not ensure that the bidders submit their true valuations. On the other hand, the SPA, in which the highest bidder pays the price offered by the second highest bidder, ensures that the bidders submit their true valuations, i.e., ensures IC, but does not maximize the revenue of the seller. Therefore, in order to ensure that both conditions of truthfulness and revenue maximization of the seller are satisfied, we design an optimal auction using the Deep Learning approach with reference to the study in [258].

4.4.2 Deep Learning Based Auction for Valuation of Cluster Heads

In this section, we illustrate the neural network architecture for the design of an optimal auction. Following the procedure in [258], we describe the neural network architecture (Fig. 4.2), which renders the design of an optimal auction. Then, we elaborate on the proposed implementation of multiple round single-item auctions for the valuation of the services of cluster heads.

By adopting the SPA scheme to determine the payment price of the winning model owner, the revenue of the cluster head is not maximized, especially when the bid of the second highest bidder is low. Thus, in order to maximize the revenue of the cluster heads, the monotonically increasing functions are applied to the bids of

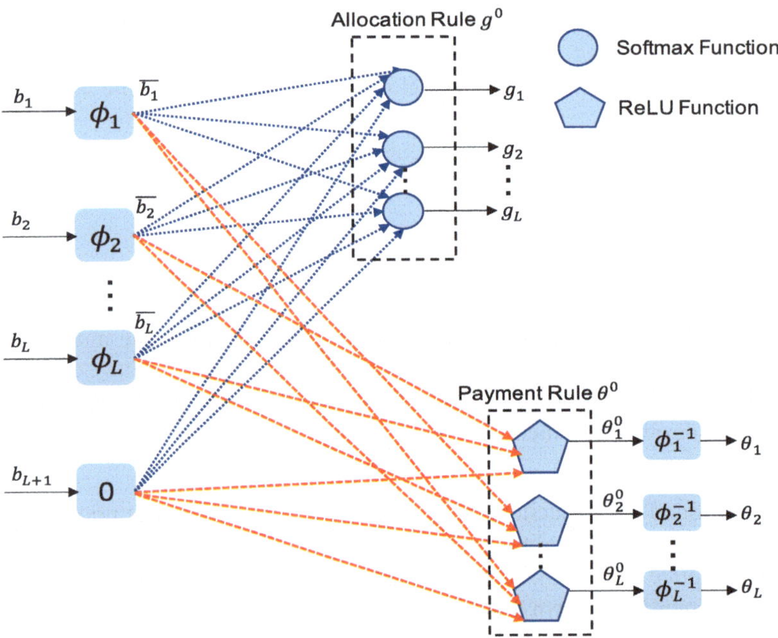

Fig. 4.2 Neural network architecture for the optimal auction

the model owners to transform the bids into transformed bids, which are used to determine the allocation and the corresponding payment of the model owners in the network. The input bids and the transformed bids of model owner l are denoted as b_l and $\bar{b}_l$, respectively. The transform function for the bids submitted by model owner l is denoted as ϕ_l. In order to determine the allocation and conditional payment of the model owners, the SPA with zero reserve price (SPA-0) is applied to transform the bids. The reserve price is the minimum price that the cluster head requires to offer its service. The SPA-0 allocation rule and the SPA-0 payment rule of the model owner l are represented by $g_l^0(\bar{\boldsymbol{b}})$ and $\theta_l^0(\bar{\boldsymbol{b}})$, respectively, where $\bar{\boldsymbol{b}}$ is the vector of the transformed bids. The SPA-0 allocation rule $g_l^0(\bar{\boldsymbol{b}})$ determines the winning model owner which has the highest bid if the bid is greater than zero. The SPA-0 payment rule $\theta_l^0(\bar{\boldsymbol{b}})$ determines the conditional payment price θ_l of model owner l by applying the inverse transform function which is represented by ϕ_l^{-1}.

Theorem 4.3 *For any set of strictly monotonically increasing function* $\{\phi_1, \ldots, \phi_L\}$, *an auction that is defined by the allocation rule* $g_l = g_l^0 \circ \phi_l$ *and the payment rule* $\theta_l = \phi_l^{-1} \circ \theta_l^0 \circ \phi_l$ *satisfies the properties of IC and IR, where* g^0 *and* θ^0 *represent the SPA-0 allocation rule and the SPA-0 payment rule, respectively [258].*

Based on the theorem, for any choices of strictly monotonically increasing transform functions, the proposed auction with the allocation rule g_l and conditional payment rule θ_l satisfy the characteristics of the optimal auction, i.e., IR and IC. Therefore, the monotone transform functions are used in the neural network to ensure the IR and IC properties of the auction. In addition to this, the cluster heads want to maximize their revenues. Based on the allocation and the conditional payment rules, the revenue of the cluster head is determined. In particular, the revenue of the cluster head is equivalent to the payment price of the winning model owner. Thus, the objective of the cluster heads is to maximize their individual revenues while fulfilling the properties of IR and IC of the optimal auction. In order to do so, the neural network architecture learns the appropriate transform functions for the optimal auction to minimize the loss function which is defined as the expectation of the negated revenue of the cluster head. The minimization of the loss function is equivalent to the maximization of the revenue of the cluster head. With this, the optimal auction design based on the neural network architecture maximizes the revenues of the cluster heads while satisfying the necessary and sufficient conditions for IC and IR.

The algorithm for the implementation of the optimal auction based on the neural network architecture is illustrated in Algorithm 4.2.

In the following, we discuss the three important functions in the neural network architecture:

1. The monotone transform function ϕ_l
2. The allocation rule g_l
3. The conditional payment rule θ_l

Algorithm 4.2 Algorithm for deep learning based optimal auction

Require: Set of cluster heads $\mathcal{J} = \{1, \ldots, j, \ldots, J\}$, bids of model owners $b^i = (b^i_1, \ldots, b^i_l, \ldots, b^i_L)$
Ensure: Revenue of the cluster heads

while $\mathcal{J} \neq \emptyset$ **do**
  Identify the cluster head the highest data coverage, D_j
  Initialization: $\mathbf{w} = [w^l_{qs}] \in \mathbb{R}_+^{I \times QS}$, $\boldsymbol{\beta} = [\beta^l_{qs}] \in \mathbb{R}^{I \times QS}$
  Deep Learning Based Optimal Auction:
  while Loss function $\hat{R}(\mathbf{w}, \boldsymbol{\beta})$ is not minimized **do**
    Compute the transformed bids $\bar{b}^i_l = \theta_l(b^i_l) = \min_{q \in \mathcal{Q}} \max_{s \in \mathcal{S}}(w^l_{qs} b_l + \beta^l_{qs})$
    Compute the allocation probabilities $g_l(\bar{\mathbf{b}}) = softmax_l(\bar{b}_1, \ldots, \bar{b}_l, \ldots, \bar{b}_{L+1}; \gamma)$
    Compute the SPA-0 payments $\theta^0_l(\bar{\mathbf{b}}) = ReLU(\max_{s \neq l} \bar{b}_s)$
    Compute the conditional payments $\theta_l = \phi^{-1}_l(\theta^0_l(\bar{\mathbf{b}}))$
    Compute the loss function $\hat{R}(\mathbf{w}, \boldsymbol{\beta})$
    Update the network parameters $\mathbf{w}$ and $\boldsymbol{\beta}$ using the SGD solver
  end while
  Update the data coverage of the winning model owner $d^{new}_l = d^{old}_l + D_j$
  Update the valuation of the winning model owner $b_l = \mu_l - d_l$
  Remove the cluster head from set $\mathcal{J}$
end while
return The revenue gain by the cluster heads

4.4.3 Monotone Transform Functions

In the auction, the valuation, i.e., bid b_l of each model owner l, is the input to the transform function ϕ_l. The transform function maps the input to its transformed bid, $\bar{b}_l = \phi_l(b_l)$. Each transform function ϕ_l is modeled as a two-layer feedforward network that consists of the min and max operators over several linear functions, as shown in Fig. 4.3. There are Q groups of S linear functions $h_{qs}(b_l) = w^l_{qs} b_l + \beta^l_{qs}$, $\mathcal{Q} = \{1, \ldots, q, \ldots, Q\}$, $\mathcal{S} = \{1, \ldots, s, \ldots, S\}$, $w^l_{qs} \in \mathbb{R}^+$, and $\beta^l_{qs} \in \mathbb{R}$ are the positive weight and bias, respectively. With these linear functions, the transform function ϕ_l of each model owner l is defined as follows:

$$\phi_l(b_l) = \min_{q \in \mathcal{Q}} \max_{s \in \mathcal{S}}(w^l_{qs} b_l + \beta^l_{qs}). \tag{4.18}$$

Based on the parameters for the forward transform function ϕ_l, the inverse function ϕ^{-1}_l can be derived as follows:

$$\phi^{-1}_l(y) = \max_{q \in \mathcal{Q}} \min_{s \in \mathcal{S}}(w^l_{qs})^{-1}(y - \beta^l_{qs}). \tag{4.19}$$

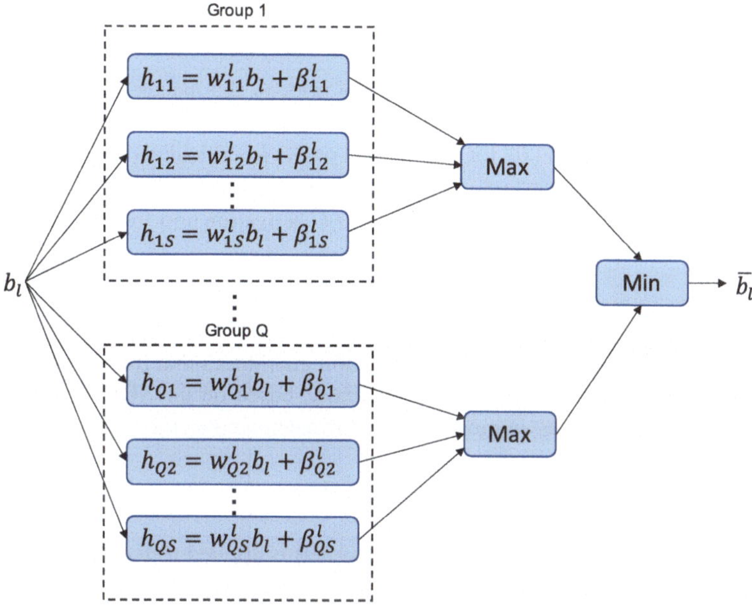

Fig. 4.3 Monotone transform functions

4.4.4 Allocation Rule

The allocation rule in the neural network architecture is based on the SPA-0 allocation rule. In particular, the data coverage D_j of the cluster head j is allocated to the model owner with the highest transformed bid if the transformed bid is more than zero. Otherwise, the cluster head does not sell its service to any model owner. In order to model the competition among the model owners, we use a *softmax* function on the vector of transformed bids $\bar{\mathbf{b}} = (\bar{b}_1, \ldots, \bar{b}_l, \ldots, \bar{b}_L)$ and the additional dummy input $\bar{b}_{L+1} = 0$ in the allocation network. The output of the *softmax* function is a vector of allocation probabilities, which is represented by $\mathbf{g} = (g_1, \ldots, g_l, \ldots, g_L)$. The *softmax* function used in the neural network architecture is defined as follows:

$$\begin{aligned} g_l(\bar{\mathbf{b}}) &= softmax_l(\bar{b}_1, \ldots, \bar{b}_l, \ldots, \bar{b}_{L+1}; \gamma) \\ &= \frac{e^{\gamma \bar{b}_l}}{\sum_{l=1}^{L+1} e^{\gamma \bar{b}_l}}, \ \gamma > 0, \forall l \in \mathcal{L}. \end{aligned} \tag{4.20}$$

The parameter γ in the *softmax* function measures the quality of the approximation where the higher the γ, the more accurate the approximation of the function. However, a better quality of approximation results in a less smooth allocation function.

4.4.5 Conditional Payment Rule

The conditional payment rule determines the price θ_l that needs to be paid by the winning model owner l. The conditional payment rule is carried out in two steps. Firstly, the SPA-0 payment θ_l^0 for each model owner l is calculated. Specifically, the SPA-0 payment θ_l^0 is the maximum of the transformed bids of other model owners and zero which is determined by using the *ReLU* activation unit function as follows:

$$\theta_l^0(\bar{\mathbf{b}}) = ReLU(\max_{s \neq l} \bar{b}_s), \ \forall l \in \mathcal{L}. \tag{4.21}$$

The $ReLU(x) = \max(x, 0)$ activation function guarantees that the SPA-0 payment of each model owner is non-negative. Secondly, based on the SPA-0 payment θ_l^0, the conditional payment θ_l of model owner l is calculated as follows:

$$\theta_l = \phi_l^{-1}(\theta_l^0(\bar{\mathbf{b}})), \tag{4.22}$$

where the inverse transform function from (4.19) is applied to the SPA-0 payment of the model owner l.

4.4.6 Neural Network Training

The aim of the neural network is to optimize the weights and biases of the linear functions in the neural network such that the loss function is minimized. In the neural network, the loss function is defined as the expectation of the negated revenue of the cluster head. The loss function of the neural network is formulated based on the inputs, i.e., the training dataset and the outputs, i.e., the allocation probabilities and the conditional payments of the model owners. The training dataset of the neural network consists of the bidders' valuation profiles of which the bidders' valuation profile i is denoted as $\mathbf{b}^i = (b_1^i, \ldots, b_L^i), \mathcal{I} = \{1, \ldots, i, \ldots, I\}$, where I is the size of the training dataset. b_l^i is the valuation of model owner l for the data coverage of cluster head is drawn from a valuation distribution function $f_B(b)$. Since the valuation b_l^i of the model owner l depends on the data coverage requirement μ_l and the current amount of data coverage d_l of the model owner, i.e., $b_l^i = \mu_l^i - d_l^i$, the distribution function $f_B(b)$ can be determined based on the distribution of the data coverage requirement, which is represented by $f\mu(\mu)$. In this chapter, we assume that the data coverage requirement of the model owners follows a uniform distribution, i.e., $\mu \sim U[\mu_{min}, \mu_{max}]$.

The parameters of the monotone transform functions, i.e., weights w_{qs}^l and biases β_{qs}^l, are the entries of matrices $\mathbf{w}$ and $\boldsymbol{\beta}$. The matrices are needed to determine the allocation probability and conditional payment of model owner l, which are represented by $g_l^{(\mathbf{w},\boldsymbol{\beta})}$ and $\theta_l^{(\mathbf{w},\boldsymbol{\beta})}$, respectively.

The objective of the training is to find the optimal weight $\mathbf{w}^*$ and bias $\boldsymbol{\beta}^*$ matrices that minimize the loss function of the neural network, i.e., the expectation of the negated revenue of the cluster head j. Specifically, the approximation of the loss function, $\hat{R}$, is defined as follows:

$$\hat{R}(\mathbf{w}, \boldsymbol{\beta}) = -\frac{1}{I}\sum_{l=1}^{I} g_l^{(\mathbf{w},\boldsymbol{\beta})}(\mathbf{b}^i)\theta_l^{(\mathbf{w},\boldsymbol{\beta})}(\mathbf{b}^i). \tag{4.23}$$

For the optimization of the loss function $\hat{R}(\mathbf{w}, \boldsymbol{\beta})$ over the parameters $(\mathbf{w}, \boldsymbol{\beta})$, a stochastic gradient descent (SGD) solver is used.

4.5 Performance Evaluation

In this section, we present the performance evaluation of the evolutionary game based cluster formation and deep learning based auction for the valuation of cluster data coverage. Unless otherwise stated, the simulation parameters are as shown in Table 4.1 [256, 259]. Note that we use the terms "cluster" and "cluster heads" interchangeably.

Table 4.1 Simulation parameters

Parameter	Values
Total number of workers in the network N	90
Number of data samples of population m D_m	[2400, 4800]
Reward pool offered in cluster α_j	[100, 300]
Fixed rewards offered in cluster R_j	80
Congestion coefficient ζ_j	[10, 20]
Computation cost c^{cmp}	0.1
Rate of strategy adaptation δ	0.001
Total Number of Model Owners, L	10
Size of Training Dataset, I	1000
Number of Groups for Linear Function Q	5
Linear Functions in Each Group S	10
Learning Rate	0.001
Quality of Approximation	1000, 2000
Range of the model owners' data coverage requirement, μ_l	$\sim U[0.5, 0.9]$, $\sim U[0, 0.4]$

4.5.1 Lower-Level Evolutionary Game

For the first part of our simulations, we analyze the lower-level evolutionary game. We consider a network that consists of 90 workers. The workers have different data quantities that follow a uniform distribution. Using the k-means algorithm, we derive 3 populations[5] of 30 workers each based on the data quantities that they possess. In the first population $m = 1$, each worker has 80 data samples. In the second population $m = 2$, each worker has 100 data samples. In the third population $m = 3$, each worker has 160 samples. Without loss of generality, the data samples that each worker owns are assumed to be characterized by the populations that they belong to. Thus, the populations are arranged in ascending order based on the data samples each worker has.

Besides, there exist 3 cluster heads in the network with each offering different reward pools α_j, as well as congestion coefficients ζ_j. Recall from Sect. 4.3.2 that a higher value of α_j indicates that a cluster offers a larger reward pool for the workers to share, whereas a higher value of ζ_j represents that a cluster head has more limited communication resources. The clusters are arranged in ascending order based on the reward pool they offer, i.e., cluster 3 offers the highest reward pool to its workers.

Accordingly, in each time period, workers in the populations choose one of the clusters to join. Then, following Algorithm 4.1, the strategy adaptation is performed and evolved such that the workers evaluate their payoffs and churn to another cluster with higher payoffs with some probabilities. Eventually, the evolutionary equilibrium is achieved.

4.5.1.1 Stability and Uniqueness of the Evolutionary Equilibrium

To demonstrate the uniqueness of the evolutionary equilibrium, i.e., the solution to the replicator dynamics defined in (4.7) from Sect. 4.3.2, we first derive the phase plane of the replicator dynamics in Fig. 4.4. For ease of exposition, population 3 is excluded initially. As such, only the first and second populations of workers are considered to choose among the three clusters to join. Figure 4.4 shows the population states of populations 1 and 2, i.e., the proportion of workers in each population that join cluster 1. We consider varying initial conditions in Fig. 4.4 and plot the corresponding dynamics. For example, for the first condition, we have 10% of the workers from both populations choose cluster 1. Clearly, despite varying initial conditions, the evolutionary equilibrium always converges to a unique solution.

To evaluate the stability of this evolutionary equilibrium, we consider that 30% and 70% of workers in populations 1 and 2, respectively, are initially allocated to cluster 1. Then, we plot the evolution of population states in Fig. 4.5. We observe

[5] For ease of exposition, we use only three populations for comparison. Our model can easily be extended to multiple populations.

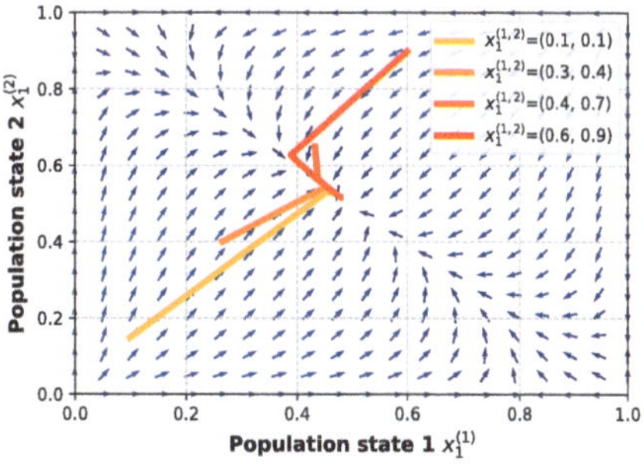

Fig. 4.4 Phase plane of the replicator dynamics

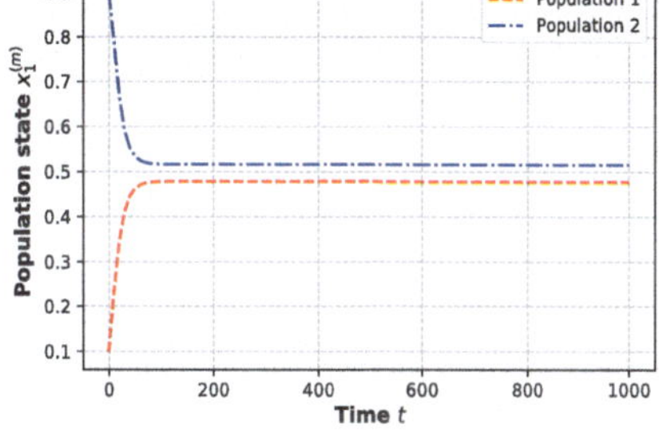

Fig. 4.5 Evolutionary equilibrium of population states for cluster 1

that the proportion of workers from population 2 joining cluster 1 declines as more workers from population 1 join the cluster. This is due to the division of rewards and congestion effects as more workers join the cluster. Eventually, the evolutionary equilibrium is reached and the population states no longer change.

Next, we consider the situation in which there are three populations and three clusters. We set the initial conditions such that a third of the workers from each population are assigned to each cluster initially. Then, we plot the evolution of utilities in Fig. 4.6. Specifically, within each population, we plot the utilities derived by workers which have chosen each of the three clusters. We observe that for each population, the utilities derived from choosing the varying clusters eventually converge with time. This implies that an evolutionary equilibrium is reached whereby at the equilibrium, the workers no longer have the incentive to adapt their cluster selection strategies. Moreover, at the equilibrium, workers that belong to population 3 derive the greatest utilities, given that they are compensated for having larger data shares in the clusters.

In Figs. 4.7, 4.8, and 4.9, we plot the evolution of population states of each population, respectively. We observe that cluster 3, which offers the highest reward pool for distribution across workers, has the largest proportion of workers from population 3. The reason is that the workers from population 3 have the largest number of data samples. Hence, they can have the largest shares of the large reward pool if they join cluster 3. In contrast, the lowest proportion of workers from population 1 joins cluster 3 because they have the lowest reward shares. However, there is an upper limit to how many workers can join cluster 3. Even though cluster 3 offers the highest reward pool, the distribution of rewards and congestion effects

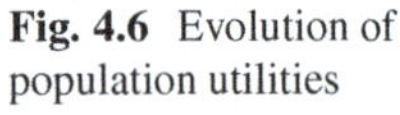

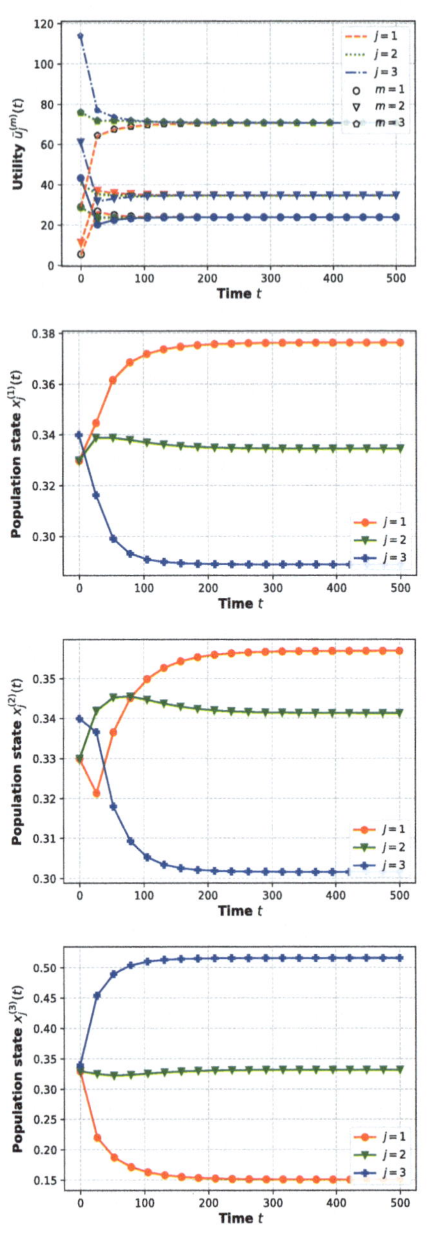

Fig. 4.6 Evolution of population utilities

Fig. 4.7 Evolution of population states for population 1

Fig. 4.8 Evolution of population states for population 2

Fig. 4.9 Evolution of population states for population 3

set in eventually when more workers have joined the cluster. Thus, workers of population 3 may also join clusters 1 and 2 as well when this occurs.

4.5.1.2 Evolutionary Equilibrium Under Varying Parameters and Conditions

In this section, we vary the simulation parameters to study the evolutionary equilibrium under varying conditions. In Fig. 4.10, we vary the learning rate of the population, which controls the speed of strategy adaptation. Naturally, when the learning rate is low, the evolutionary equilibrium can only be reached after a longer time period. However, we can observe that the stability of the evolutionary equilibrium is not compromised. The speed of convergence depends on how fast the workers can observe and adapt their strategies, e.g., they have more accurate information about the system.

In Fig. 4.11, we vary the reward pool offered by cluster 1 between [50, 350] while keeping the reward pool for clusters 2 and 3 constant at 200 and 300, respectively. Then, we plot the changes in data coverage of each cluster, i.e., how much data coverage a cluster has as a result of worker contribution. Naturally, the data coverage of cluster 1 increases with an increment of the reward pool. The reason is that the cluster is more attractive to workers as shareable rewards increase. We further note that at the points $\alpha_1 = 200$ and $\alpha_1 = 300$, the data coverage for cluster 1 is identical to those of clusters 2 and 3, respectively. The reason is that at these points, cluster 1 is identical to the corresponding clusters, and thus, workers are indifferent between choosing the cluster to join.

In Fig. 4.12, we vary the congestion coefficient of cluster 1 between [2, 18] while keeping the other clusters' constant. Then, we plot the changes in data coverage of each cluster. We observe that as the congestion coefficient increases, the cluster

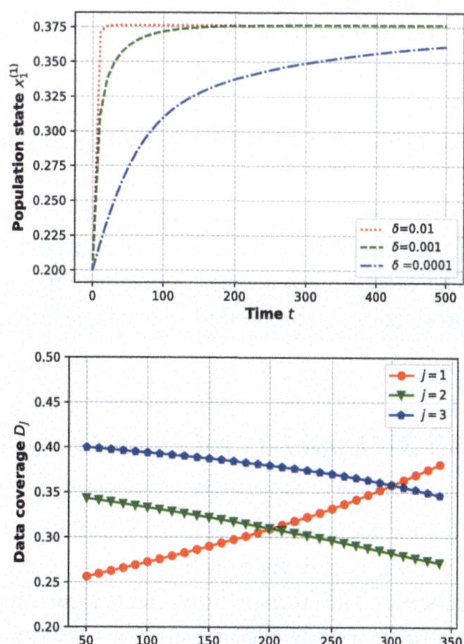

Fig. 4.10 Evolutionary dynamics under different learning rates

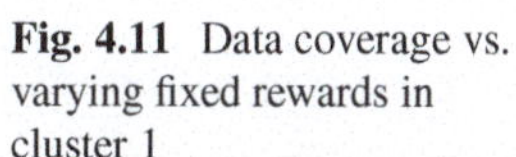

Fig. 4.11 Data coverage vs. varying fixed rewards in cluster 1

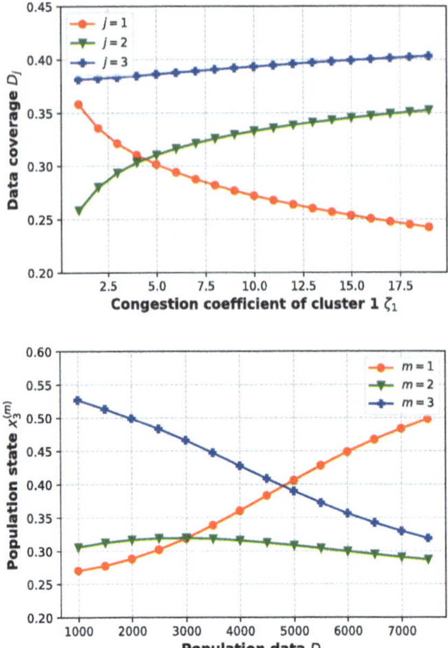

Fig. 4.12 Data coverage vs. varying congestion coefficient in cluster 1

Fig. 4.13 Population states in cluster 3 vs. varying population data for population 1

has lower data coverage. Instead, the workers that used to join cluster 1 adapt their strategies and churn to clusters 2 and 3. This is given that with a large congestion coefficient, the cluster head has limited communication resources and can no longer support as many workers without them having to incur higher communication costs, e.g., due to device interference.

In Fig. 4.13, we vary the data quantities of workers in population 1 while keeping the other populations constant. Then, we plot the population states of all populations with respect to participation in cluster 3. Specifically, the figure shows the proportion of workers from each population which has joined cluster 3 as the data quantities of population 1 vary. Clearly, as the data owned by workers in population 1 increases, more workers from population 1 are able to join cluster 3, which is the cluster with the largest reward pool as mentioned in Sect. 4.5.1.1. The reason is that with more data, the workers in the population can gain a larger proportion of the pooled rewards, relative to that of workers from other populations.

4.5.2 Upper-Level Deep Learning Based Auction

In this section, we perform simulations to evaluate the performance of the deep learning based auction. For comparison, the classic SPA is chosen as the baseline scheme. The TensorFlow Deep Learning library is used to implement the optimal auction design. We consider a network with the formations of 3 clusters and 10 model owners. We first evaluate the performance of the deep learning based auction against the traditional SPA. Then, we proceed to study the impacts of (a) data coverage of the cluster heads, (b) model owners with varying distribution for data coverage requirement, and (c) different quality of approximation, on the revenues of the cluster heads.

4.5.2.1 Evaluation of the Deep Learning Based Auction

From Figs. 4.14, 4.15, and 4.16, we observe that the revenue of the cluster heads determined by the deep learning based auction is consistently higher than that of the conventional SPA scheme. The reason is that the SPA scheme guarantees IC but does not guarantee that the revenue of the seller is maximized. While preserving the properties of IC and IR of the traditional auction, the deep learning based auction maximizes the revenue earned by the cluster heads by providing their services to the model owners that value their services the most.

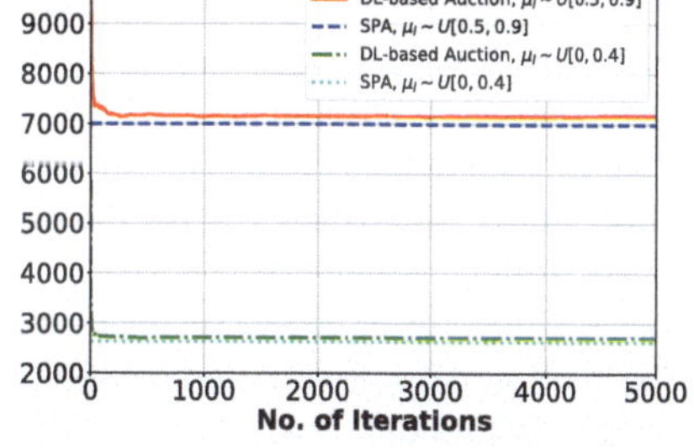

Fig. 4.14 Revenue of cluster head 1 under different distribution of model owners

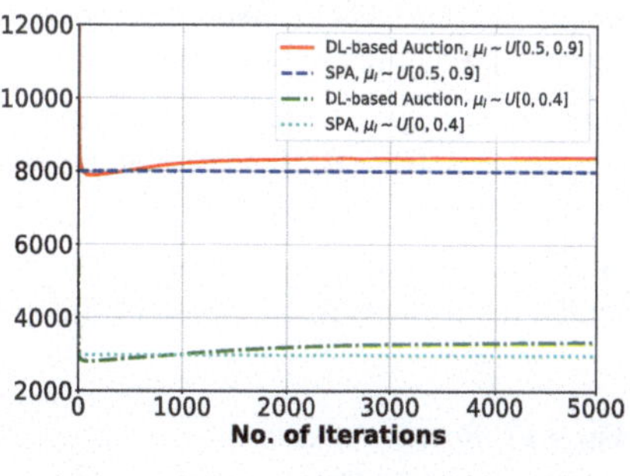

Fig. 4.15 Revenue of cluster head 2 under different distribution of model owners

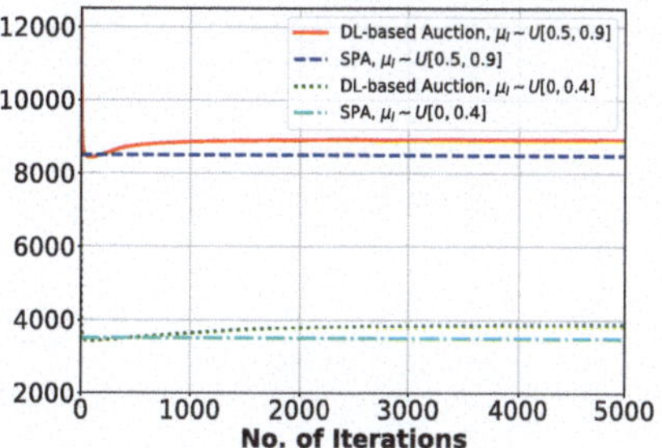

Fig. 4.16 Revenue of cluster head 3 under different distribution of model owners

We examine the impacts on the cluster heads' revenues when they are presented with model owners with data coverage requirements of different distribution ranges. Specifically, model owners can take on two distribution ranges, i.e., $\mu_l \sim U[0, 0.4]$ and $\mu_l \sim U[0.5, 0.9]$. In Fig. 4.14, cluster head 1 that has a total data coverage of 0.26 earns a revenue of 2724 when model owners have data coverage requirements that range between 0 and 0.4. On the other hand, when model owners have a higher range of data coverage requirements, i.e., between 0.5 and 0.9, cluster head 1 earns a higher revenue of 7182. Since the model owners with higher data coverage requirements value the cluster head more, they have more incentive to pay a higher price which results in the higher revenue earned by the cluster head. Similar trends are observed for cluster head 2 and cluster head 3 in Figs. 4.15 and 4.16, respectively.

The revenues of the cluster heads are affected by the amount of data coverage. In Fig. 4.17, we observe that when the data coverage of the cluster head is higher, the revenue earned is also higher. In particular, when the model owners have high requirement for data coverage, i.e., $\mu_l \sim U[0.5, 0.9]$, cluster head 3 with the highest data coverage of 0.4 earns a revenue of 8944, whereas cluster head 1 with the lowest data coverage of 0.26 earns a revenue of 7182. This is due to the fact that the cluster head with the highest data coverage is allowed to conduct auction for its services first. Hence, it will be able to offer its service to the model owner with the highest data coverage requirement in which the model owner is willing to pay the highest price as compared to other model owners in the network. Intuitively, this serves to compensate the cluster head for the higher rewards expense it incurs.

To further evaluate the performance of the deep learning based auction, we consider the cluster heads' revenues under the different quality of approximation, γ. The quality of approximation is used in the *softmax* function in the determination of the winning model owner. We consider two values for the quality of approximation, i.e., 1000 and 2000. We observe in Figs. 4.18, 4.19, and 4.20, the revenue of the cluster head increases slightly when the quality of approximation is higher. Specifically, given the model owners with a high requirement for data coverage, i.e., $\mu_l \sim U[0.5, 0.9]$, the revenues of cluster head 3 are 8944 and 8969 when the values of γ are 1000 and 2000, respectively. This follows that with a greater value of approximation quality, the neural network is able to solve the optimization problem which maximizes the revenue of the cluster heads.

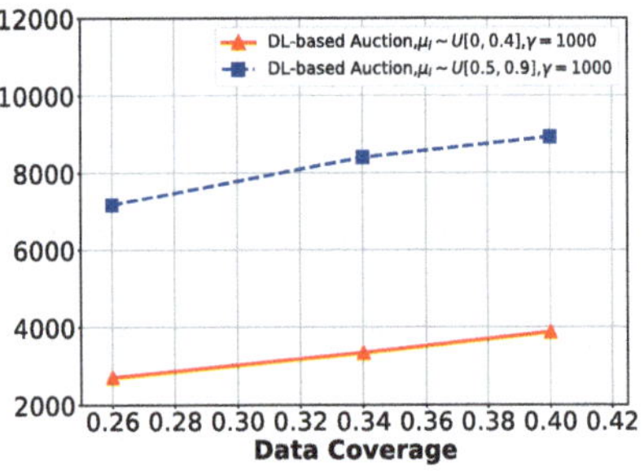

Fig. 4.17 Revenue vs data coverage of cluster heads

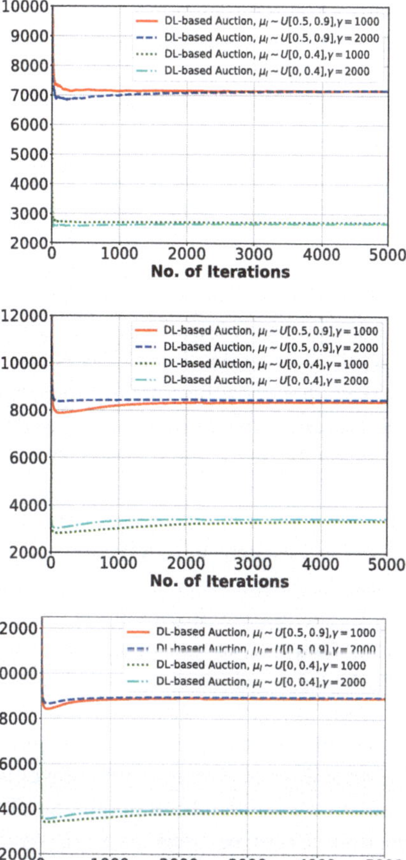

Fig. 4.18 Revenue of cluster head 1 under different approximation qualities

Fig. 4.19 Revenue of cluster head 2 under different approximation qualities

Fig. 4.20 Revenue of cluster head 3 under different approximation qualities

4.6 Conclusion and Chapter Discussion

In this chapter, we proposed a resource allocation and incentive mechanism design framework for HFL. We considered a two-level problem and leveraged the evolutionary game theory to derive the equilibrium solution for the cluster selection phase. Then, we introduced a deep learning based auction mechanism to value the cluster head's services. The performance evaluation shows the uniqueness and stability of the evolutionary equilibrium, as well as the revenue maximizing property of the auction mechanism.

Following our discussions in Chaps. 2 and 3, this chapter differs in that there is no monopoly for the FL model owner. In other words, there is competition among the FL model owners. To model the competition, we have utilized an auction mechanism in this chapter. Moreover, while Chaps. 2 and 3 assume the static network composition, we have considered the case of dynamic network formation in a self-organized HFL network in this chapter. As such, this challenge is solved using the evolutionary game theory aptly. On top of that, the convergence of the algorithm is proven and shown numerically. In the following, we discuss the drawbacks of the chapter and the potential future works.

Firstly, in this chapter, we have assumed that the workers are only able to join a single model owner. The practical reason is that at each instance, the worker is only able to support one instance of model training. Moreover, given that the model owners may be competing and that the model itself has value, it is unlikely that the workers will be allowed to join multiple instances of training. However, some workers may deviate from this assumption and join other instances of model training at separate time slots. To circumvent this shortfall, one method is to maintain the workers' reputations and memberships on a reputation blockchain [41]. In this case, workers that cheat by joining more than one instance of training will be penalized and prevented from joining other model owners.

Secondly, we have assumed that the cluster head's reward pools are static in this chapter. This assumption is not unrealistic, given that in reality, a cluster head may have fixed budget constraints. To further account for the cases in which the cluster heads may have varying budget constraints, we can explore the use of a Stackelberg differential game. As an illustration, in the lower-level game, the cluster head association strategies of the workers can still be modeled using an evolutionary game. However, in the upper-level game, a Stackelberg differential game [260] can be adopted in which the cluster heads decide an optimal reward scheme given the dynamic association strategies of the workers in the lower level.

Thirdly, the model owners only purchase the intermediate model from the cluster heads in this chapter. In reality, the cluster heads' communication and computation resources have also been purchased. To cover this resource expense, a single-item auction cannot be used. Instead, we can consider a combinatorial auction with varying valuation profiles. The combinatorial auction can be solved in two phases. Firstly, the winners' determination problem can be solved using a Graph Neural Network based approach [261]. Secondly, the pricing problem can be solved using our approach discussed in this chapter.

Fourthly, in this chapter, we have made the assumption that a worker has a comparatively small subset of data as compared to other workers. This may be the case, e.g., for IoT collected data in a large-scale network of users. The utility gained from training the derived model is therefore not incorporated within this chapter. However, for some cases, the utility gained from cooperatively training a better model may be significant. One such example is FL for finance, in which a coalition of financial institutions may band together to train a model. If an institution does not take part, it may not be allowed to use the model. In such a case, the utility may therefore be modeled differently, and the focus is on contributions to receive a common model, rather than an individual payment.

Finally, in a large-scale HFL network, we recognize that it is difficult to track the malicious updates by the workers. For example, the model poisoning attacks by malicious workers may corrupt the inference accuracy of the derived intermediate model and, in turn, affect the global model that the FL model owner attempts to build. As such, consensus mechanisms such as the proof of verification [262] can be considered. This can be implemented in tandem with the reputation blockchain [41] that we have discussed previously.

Chapter 5
Conclusion and Future Works

In this book, we began with a comprehensive survey of FL enabled edge intelligence in Chap. 1. In the chapter, we highlighted the enabling role that FL has in future edge intelligence networks as a tool to facilitate and empower scalable AI at the edge. Then, we discussed the drawbacks and limitations of FL. In particular, the challenges of communication inefficiency, resource misallocation, and privacy and security intrusions are crucial issues to be solved before the scalable implementation of FL.

Given the wide scope of these pressing challenges, we began the discussion of technical works in Chap. 2 by addressing the incentive mechanism design problem. Recognizing the limitations of vanilla contract optimization, we formulate a multi-dimensional contract and matching optimization framework. Specifically, we consider a scenario of UAV and AI empowered edge intelligence networks. The UAVs are characterized by multiple sources of resource heterogeneity. As such, the multi-dimensional contract is useful to account for this aspect.

In Chap. 3, we discuss that besides utilizing UAVs as aerial sensing devices, there is potential for UAVs to bridge the gap between the terrestrial data owners and cloud by aiding in intermediate aggregation. In light of the hardware constraints of UAVs, we introduce a joint auction-coalition framework for the optimal formation of UAV swarm and pricing of UAV services.

However, the works presented in Chaps. 2 and 3 do not capture the self-organizing and dynamic aspect of an FL network. We propose a hierarchical resource allocation framework for communication efficient hierarchical FL in Chap. 4. In the lower level, we devise an evolutionary game model to derive the equilibrium composition of the clusters. In the upper level, to derive the allocation of cluster head to the model owner, as well as the optimal pricing of the services of the cluster head by the competitive model owners, we adopt a deep learning-based auction mechanism.

As a conclusion to the book, we discuss the challenges and research directions in deploying FL at scale to be discussed as follows.

W. Y. B. Lim et al., *Federated Learning Over Wireless Edge Networks*, Wireless Networks, https://doi.org/10.1007/978-3-031-07838-5_5

1. *Dropped participants:* The approaches discussed in Sect. 1.4, e.g., [88], [91], and [92], propose new algorithms for participant selection and resource allocation to address the training bottleneck and resource heterogeneity. In these approaches, the wireless connections of participants are assumed to be always available. However, in practice, participating mobile devices may go offline and can drop out from the FL system due to connectivity or energy constraints. A large number of dropped devices from the training participation can significantly degrade the performance [21], e.g., accuracy and convergence speed, of the FL system. New FL algorithms need to be robust to device drop out in the networks and anticipate the scenarios in which only a small number of participants are left connected to participate in a training round. One potential solution is that the FL model owner provides free dedicated/special connections, e.g., cellular connections, as an incentive to the participants to avoid drop out.
2. *Privacy concerns:* FL is able to protect the data privacy of each participants since the raw data of the participant is not exposed to a third-party server, with just the model parameters exchanged with the FL server. However, as specified in [121], [122], and [123], communicating the model updates during the training process can still reveal sensitive information to an adversary or a third party. The current approaches propose security solutions such as DP, e.g., [20], [128], and [263], and collaborative training, e.g., [129] and [130]. However, the adoption of these approaches sacrifices the performance, i.e., model accuracy. They also require significant computation on participating mobile devices. Thus, the tradeoff between privacy guarantee and system performance has to be well balanced when implementing the FL system.
3. *Unlabeled data:* It is important to note that the approaches reviewed in the survey are proposed for supervised learning tasks. This means that the approaches assume that labels exist for all of the data in the federated network. However, in practice, the data generated in the network may be unlabeled or mislabeled [264]. This poses a big challenge to the server to find participants with appropriate data for model training. Tackling this challenge may require the challenges of scalability, heterogeneity, and privacy in the FL systems to be addressed. One possible solution is to enable mobile devices to construct their labeled data by learning the "labeled data" from each other. Emerging studies have also considered the use of semi-supervised learning-inspired techniques [265].
4. *Interference among mobile devices:* The existing resource allocation approaches, e.g., [88] and [92], address the participant selection based on the resource states of their mobile devices. In fact, these mobile devices may be geographically close to each other, i.e., in the same cell. This introduces an interference issue when they update local models to the server. As such, channel allocation policies may need to be combined with the resource allocation approaches to address the interference issue. While studies in [99], [101], and [102] consider multi-access schemes and over-the-air computation, it remains to be seen if such approaches are scalable, i.e., able to support a large federation

of many participants. To this end, data-driven learning-based solutions, e.g., federated DRL, can be considered to model the dynamic environment of mobile edge networks and make optimized decisions.

5. *Comparisons with other distributed learning methods:* Following the increased scrutiny on data privacy, there has been a growing effort on developing new privacy preserving distributed learning algorithms. One study proposes *split learning* [266], which also enables collaborative ML without requiring the exchange of raw data with an external server. In split learning, each participant first trains the neural network up to a cut layer. Then, the outputs from training are transmitted to an external server that completes the other layers of training. The resultant gradients are then backpropagated up to the cut layer and eventually returned to the participants to complete the local training. In contrast, FL typically involves the communication of full model parameters. The authors in [267] conduct an empirical comparison between the communication efficiencies of split learning and FL. The simulation results show that split learning performs well when the model size involved is larger, or when there are more participants involved, since the participants do not have to transmit the weights to an aggregating server. However, FL is much easier to implement since the participants and FL server are running the same global model, i.e., the FL server is just in charge of aggregation and thus FL can work with one of the participants serving as the master node. As such, more research efforts can be directed toward guiding system administrators to make an informed decision as to which scenario warrants the use of either learning method.
6. *Further studies on learning convergence:* One of the essential considerations of FL is the convergence of the algorithm. FL finds weights to minimize the global model aggregation. This is actually a distributed optimization problem, and the convergence is not always guaranteed. Theoretical analysis and evaluations on the convergence bounds of the gradient descent based FL for convex and non-convex loss functions are important research directions. While existing studies have covered this topic, many of the guarantees are limited to restrictions, e.g., the convexity of the loss function.
7. *Usage of tools to quantify statistical heterogeneity:* Mobile devices typically generate and collect data in a non-IID manner across the network. Moreover, the number of data samples among the mobile devices may vary significantly. To improve the convergence of FL algorithm, the statistical heterogeneity of the data needs to be quantified. Recent works, e.g., [268], have developed tools for quantifying statistical heterogeneity through metrics such as local dissimilarity. However, these metrics cannot be easily calculated over the federated network before training begins. The importance of these metrics motivates future directions such as the development of efficient algorithms to quickly determine the level of heterogeneity in federated networks.
8. *Combined algorithms for communication reduction:* Currently, there are three common techniques of communication reduction in FL as discussed in Sect. 1.3. It is important to study how these techniques can be combined

with each other to improve the performance further. For example, the model compression technique can be combined with the edge server-assisted FL. The combination is able to significantly reduce the size of model updates, as well as the instances of communication with the FL server. However, the feasibility of this combination has not been explored. In addition, the tradeoff between accuracy and communication overhead for the combination technique needs to be further evaluated. In particular, for simulation results we discuss in Sect. 1.3, the accuracy-communication cost reduction tradeoff is difficult to manage since it varies for different settings, e.g., data distribution, quantity, number of edge servers, and number of participants.

9. *Cooperative mobile crowd ML:* In the existing approaches, mobile devices need to communicate with the server directly and this may increase the energy consumption. In fact, mobile devices nearby can be grouped in a cluster, and the model downloading/uploading between the server and the mobile devices can be facilitated by a "cluster head" that serves as a relay node [269]. The model exchange between the mobile devices and the cluster head can then be done in Device-to-Device (D2D) connections. Such a model can improve the energy efficiency significantly. Efficient coordination schemes for the cluster head can thus be designed to further improve the energy efficiency of an FL system.
10. *Applications of FL:* Given the advantages of guaranteeing data privacy, FL has an increasingly important role to play in many applications, e.g., healthcare, finance, and transport systems. For most current studies on FL applications, the focus mainly lies in the federated training of the learning model, with the implementation challenges neglected. For future studies on the applications of FL, besides a need to consider the aforementioned issues in the survey, i.e., communication costs, resource allocation, and privacy and security, there is also a need to consider the specific issues related to the system model in which FL will be adopted in. For example, for delay critical applications, e.g., in vehicular networks, there will be more emphasis on training efficiency and less on energy expense.

References

1. Y. LeCun, L. Bottou, Y. Bengio, P. Haffner *et al.*, "Gradient-based learning applied to document recognition," *Proceedings of the IEEE*, vol. 86, no. 11, pp. 2278–2324, 1998.
2. X. Wang, Y. Han, C. Wang, Q. Zhao, X. Chen, and M. Chen, "In-edge ai: Intelligentizing mobile edge computing, caching and communication by federated learning," *arXiv preprint arXiv:1809.07857*, 2018.
3. R. Pryss, M. Reichert, J. Herrmann, B. Langguth, and W. Schlee, "Mobile crowd sensing in clinical and psychological trials–a case study," in *2015 IEEE 28th International Symposium on Computer-Based Medical Systems*. IEEE, 2015, pp. 23–24.
4. R. K. Ganti, F. Ye, and H. Lei, "Mobile crowdsensing: current state and future challenges," *IEEE Communications Magazine*, vol. 49, no. 11, pp. 32–39, 2011.
5. Y. LeCun, Y. Bengio, and G. Hinton, "Deep learning," *Nature*, vol. 521, no. 7553, p. 436, 2015.
6. D. Oletic and V. Bilas, "Design of sensor node for air quality crowdsensing," in *2015 IEEE Sensors Applications Symposium (SAS)*. IEEE, 2015, pp. 1–5.
7. Y. Jing, B. Guo, Z. Wang, V. O. Li, J. C. Lam, and Z. Yu, "Crowdtracker: Optimized urban moving object tracking using mobile crowd sensing," *IEEE Internet of Things Journal*, vol. 5, no. 5, pp. 3452–3463, 2017.
8. H.-J. Hong, C.-L. Fan, Y.-C. Lin, and C.-H. Hsu, "Optimizing cloud-based video crowdsensing," *IEEE Internet of Things Journal*, vol. 3, no. 3, pp. 299–313, 2016.
9. W. He, G. Yan, and L. Da Xu, "Developing vehicular data cloud services in the IoT environment," *IEEE Transactions on Industrial Informatics*, vol. 10, no. 2, pp. 1587–1595, 2014.
10. P. Li, J. Li, Z. Huang, T. Li, C.-Z. Gao, S.-M. Yiu, and K. Chen, "Multi-key privacy-preserving deep learning in cloud computing," *Future Generation Computer Systems*, vol. 74, pp. 76–85, 2017.
11. B. Custers, A. Sears, F. Dechesne, I. Georgieva, T. Tani, and S. van der Hof, *EU Personal Data Protection in Policy and Practice*. Springer, 2019.
12. B. M. Gaff, H. E. Sussman, and J. Geetter, "Privacy and big data," *Computer*, vol. 47, no. 6, pp. 7–9, 2014.
13. Y. Mao, C. You, J. Zhang, K. Huang, and K. B. Letaief, "A survey on mobile edge computing: The communication perspective," *IEEE Communications Surveys & Tutorials*, vol. 19, no. 4, pp. 2322–2358, 2017.
14. G. Ananthanarayanan, P. Bahl, P. Bodík, K. Chintalapudi, M. Philipose, L. Ravindranath, and S. Sinha, "Real-time video analytics: The killer app for edge computing," *Computer*, vol. 50, no. 10, pp. 58–67, 2017.

W. Y. B. Lim et al., *Federated Learning Over Wireless Edge Networks*, Wireless Networks, https://doi.org/10.1007/978-3-031-07838-5

15. H. Li, K. Ota, and M. Dong, "Learning IoT in edge: Deep learning for the internet of things with edge computing," *IEEE Network*, vol. 32, no. 1, pp. 96–101, 2018.
16. X. Wang, Y. Han, V. C. Leung, D. Niyato, X. Yan, and X. Chen, "Convergence of edge computing and deep learning: A comprehensive survey," *IEEE Communications Surveys & Tutorials*, vol. 22, no. 2, pp. 869–904, 2020.
17. W. Shi, J. Cao, Q. Zhang, Y. Li, and L. Xu, "Edge computing: Vision and challenges," *IEEE Internet of Things Journal*, vol. 3, no. 5, pp. 637–646, 2016.
18. X. Chen, L. Jiao, W. Li, and X. Fu, "Efficient multi-user computation offloading for mobile-edge cloud computing," *IEEE/ACM Transactions on Networking*, vol. 24, no. 5, pp. 2795–2808, 2015.
19. P. Mach and Z. Becvar, "Mobile edge computing: A survey on architecture and computation offloading," *IEEE Communications Surveys & Tutorials*, vol. 19, no. 3, pp. 1628–1656, 2017.
20. M. Abadi, A. Chu, I. Goodfellow, H. B. McMahan, I. Mironov, K. Talwar, and L. Zhang, "Deep learning with differential privacy," in *Proceedings of the 2016 ACM SIGSAC Conference on Computer and Communications Security*. ACM, 2016, pp. 308–318.
21. B. McMahan, E. Moore, D. Ramage, S. Hampson, and B. A. y Arcas, "Communication-efficient learning of deep networks from decentralized data," in *Artificial intelligence and statistics*. PMLR, 2017, pp. 1273–1282.
22. F. Cicirelli, A. Guerrieri, G. Spezzano, A. Vinci, O. Briante, A. Iera, and G. Ruggeri, "Edge computing and social internet of things for large-scale smart environments development," *IEEE Internet of Things Journal*, vol. 5, no. 4, pp. 2557–2571, 2017.
23. A. Hard, K. Rao, R. Mathews, F. Beaufays, S. Augenstein, H. Eichner, C. Kiddon, and D. Ramage, "Federated learning for mobile keyboard prediction," *arXiv preprint arXiv:1811.03604*, 2018.
24. T. S. Brisimi, R. Chen, T. Mela, A. Olshevsky, I. C. Paschalidis, and W. Shi, "Federated learning of predictive models from federated electronic health records," *International journal of medical informatics*, vol. 112, pp. 59–67, 2018.
25. K. Powell, "Nvidia clara federated learning to deliver ai to hospitals while protecting patient data," *https://blogs.nvidia.com/blog/2019/12/01/clara-federated-learning/*, 2019.
26. D. Verma, S. Julier, and G. Cirincione, "Federated AI for building AI solutions across multiple agencies," *arXiv preprint arXiv:1809.10036*, 2018.
27. O. Simeone, "A very brief introduction to machine learning with applications to communication systems," *IEEE Transactions on Cognitive Communications and Networking*, vol. 4, no. 4, pp. 648–664, 2018.
28. C. Zhang, P. Patras, and H. Haddadi, "Deep learning in mobile and wireless networking: A survey," *IEEE Communications Surveys & Tutorials*, 2019.
29. N. C. Luong, D. T. Hoang, S. Gong, D. Niyato, P. Wang, Y.-C. Liang, and D. I. Kim, "Applications of deep reinforcement learning in communications and networking: A survey," *IEEE Communications Surveys & Tutorials*, 2019.
30. K. Hamidouche, A. T. Z. Kasgari, W. Saad, M. Bennis, and M. Debbah, "Collaborative artificial intelligence (AI) for user-cell association in ultra-dense cellular systems," in *2018 IEEE International Conference on Communications Workshops (ICC Workshops)*. IEEE, 2018, pp. 1–6.
31. S. Samarakoon, M. Bennis, W. Saad, and M. Debbah, "Distributed federated learning for ultra-reliable low-latency vehicular communications," *IEEE Transactions on Communications*, vol. 68, no. 2, pp. 1146–1159, 2019.
32. Q. Yang, Y. Liu, T. Chen, and Y. Tong, "Federated machine learning: Concept and applications," *ACM Transactions on Intelligent Systems and Technology (TIST)*, vol. 10, no. 2, p. 12, 2019.
33. S. Niknam, H. S. Dhillon, and J. H. Reed, "Federated learning for wireless communications: Motivation, opportunities, and challenges," *IEEE Communications Magazine*, vol. 58, no. 6, pp. 46–51, 2020.
34. T. Li, A. K. Sahu, A. Talwalkar, and V. Smith, "Federated learning: Challenges, methods, and future directions," *IEEE Signal Processing Magazine*, vol. 37, no. 3, pp. 50–60, 2020.

35. Z. Zhou, X. Chen, E. Li, L. Zeng, K. Luo, and J. Zhang, "Edge intelligence: Paving the last mile of artificial intelligence with edge computing," *Proceedings of the IEEE*, vol. 107, no. 8, pp. 1738–1762, 2019.
36. L. Cui, S. Yang, F. Chen, Z. Ming, N. Lu, and J. Qin, "A survey on application of machine learning for internet of things," *International Journal of Machine Learning and Cybernetics*, vol. 9, no. 8, pp. 1399–1417, 2018.
37. K. Kumar, J. Liu, Y.-H. Lu, and B. Bhargava, "A survey of computation offloading for mobile systems," *Mobile Networks and Applications*, vol. 18, no. 1, pp. 129–140, 2013.
38. N. Abbas, Y. Zhang, A. Taherkordi, and T. Skeie, "Mobile edge computing: A survey," *IEEE Internet of Things Journal*, vol. 5, no. 1, pp. 450–465, 2017.
39. S. Wang, X. Zhang, Y. Zhang, L. Wang, J. Yang, and W. Wang, "A survey on mobile edge networks: Convergence of computing, caching and communications," *IEEE Access*, vol. 5, pp. 6757–6779, 2017.
40. J. Yao, T. Han, and N. Ansari, "On mobile edge caching," *IEEE Communications Surveys & Tutorials*, vol. 21, no. 3, pp. 2525–2553, 2019.
41. J. Kang, Z. Xiong, D. Niyato, S. Xie, and J. Zhang, "Incentive mechanism for reliable federated learning: A joint optimization approach to combining reputation and contract theory," *IEEE Internet of Things Journal*, vol. 6, no. 6, pp. 10 700–10 714, 2019.
42. C. J. Burges, "A tutorial on support vector machines for pattern recognition," *Data mining and knowledge discovery*, vol. 2, no. 2, pp. 121–167, 1998.
43. J. A. Cornell, *Classical and modern regression with applications*. Taylor & Francis, 1987.
44. S. Wang, T. Tuor, T. Salonidis, K. K. Leung, C. Makaya, T. He, and K. Chan, "Adaptive federated learning in resource constrained edge computing systems," *IEEE Journal on Selected Areas in Communications*, vol. 37, no. 6, pp. 1205–1221, 2019.
45. T. Lin, S. U. Stich, K. K. Patel, and M. Jaggi, "Don't use large mini batches, use local sgd," *arXiv preprint arXiv:1808.07217*, 2018.
46. Y. Zhao, M. Li, L. Lai, N. Suda, D. Civin, and V. Chandra, "Federated learning with non-iid data," *arXiv preprint arXiv:1806.00582*, 2018.
47. A. Krizhevsky, G. Hinton *et al.*, "Learning multiple layers of features from tiny images," Citeseer, Tech. Rep., 2009.
48. Y. Rubner, C. Tomasi, and L. J. Guibas, "The earth mover's distance as a metric for image retrieval," *International journal of computer vision*, vol. 40, no. 2, pp. 99–121, 2000.
49. M. Duan, "Astraea: Self-balancing federated learning for improving classification accuracy of mobile deep learning applications," *arXiv preprint arXiv:1907.01132*, 2019.
50. S. C. Wong, A. Gatt, V. Stamatescu, and M. D. McDonnell, "Understanding data augmentation for classification: when to warp?" in *2016 international conference on digital image computing: techniques and applications (DICTA)*. IEEE, 2016, pp. 1–6.
51. J. M. Joyce, "Kullback-Leibler divergence," *International encyclopedia of statistical science*, pp. 720–722, 2011.
52. M. Jaggi, V. Smith, M. Takác, J. Terhorst, S. Krishnan, T. Hofmann, and M. I. Jordan, "Communication-efficient distributed dual coordinate ascent," in *Advances in neural information processing systems*, 2014, pp. 3068–3076.
53. V. Smith, C.-K. Chiang, M. Sanjabi, and A. S. Talwalkar, "Federated multi-task learning," in *Advances in Neural Information Processing Systems*, 2017, pp. 4424–4434.
54. J. C. Bezdek and R. J. Hathaway, "Convergence of alternating optimization," *Neural, Parallel & Scientific Computations*, vol. 11, no. 4, pp. 351–368, 2003.
55. M. G. Arivazhagan, V. Aggarwal, A. K. Singh, and S. Choudhary, "Federated learning with personalization layers," *arXiv preprint arXiv:1912.00818*, 2019.
56. J. Ren, X. Shen, Z. Lin, R. Mech, and D. J. Foran, "Personalized image aesthetics," in *Proceedings of the IEEE International Conference on Computer Vision*, 2017, pp. 638–647.
57. X. Li, K. Huang, W. Yang, S. Wang, and Z. Zhang, "On the convergence of fedavg on non-iid data," *arXiv preprint arXiv:1907.02189*, 2019.
58. L. Huang, Y. Yin, Z. Fu, S. Zhang, H. Deng, and D. Liu, "Loadaboost: Loss-based adaboost federated machine learning on medical data," *arXiv preprint arXiv:1811.12629*, 2018.

59. M. Abadi, P. Barham, J. Chen, Z. Chen, A. Davis, J. Dean, M. Devin, S. Ghemawat, G. Irving, M. Isard *et al.*, "Tensorflow: A system for large-scale machine learning," in *12th {USENIX} symposium on operating systems design and implementation ({OSDI} 16)*, 2016, pp. 265–283.
60. T. Ryffel, A. Trask, M. Dahl, B. Wagner, J. Mancuso, D. Rueckert, and J. Passerat-Palmbach, "A generic framework for privacy preserving deep learning," *arXiv preprint arXiv:1811.04017*, 2018.
61. S. Caldas, P. Wu, T. Li, J. Konečný, H. B. McMahan, V. Smith, and A. Talwalkar, "Leaf: A benchmark for federated settings," *arXiv preprint arXiv:1812.01097*, 2018.
62. A. Go, R. Bhayani, and L. Huang, "Sentiment140," *Site Functionality, 2013c. URL http://help.sentiment140.com/site-functionality.Abrufam*, vol. 20, 2016.
63. F. Zheng, K. Li, J. Tian, X. Xiang *et al.*, "A vertical federated learning method for interpretable scorecard and its application in credit scoring," *arXiv preprint arXiv:2009.06218*, 2020.
64. J. Konečný, H. B. McMahan, D. Ramage, and P. Richtárik, "Federated optimization: Distributed machine learning for on-device intelligence," *arXiv preprint arXiv:1610.02527*, 2016.
65. J. Konečný, H. B. McMahan, F. X. Yu, P. Richtárik, A. T. Suresh, and D. Bacon, "Federated learning: Strategies for improving communication efficiency," *arXiv preprint arXiv:1610.05492*, 2016.
66. K. He, X. Zhang, S. Ren, and J. Sun, "Deep residual learning for image recognition," in *Proceedings of the IEEE conference on computer vision and pattern recognition*, 2016, pp. 770–778.
67. W. Luping, W. Wei, and L. Bo, "Cmfl: Mitigating communication overhead for federated learning," in *2019 IEEE 39th International Conference on Distributed Computing Systems (ICDCS)*. IEEE, 2019, pp. 954–964.
68. H. Wang, S. Sievert, S. Liu, Z. Charles, D. Papailiopoulos, and S. Wright, "Atomo: Communication-efficient learning via atomic sparsification," in *Advances in Neural Information Processing Systems*, 2018, pp. 9850–9861.
69. S. U. Stich, J.-B. Cordonnier, and M. Jaggi, "Sparsified sgd with memory," in *Advances in Neural Information Processing Systems*, 2018, pp. 4447–4458.
70. S. Caldas, J. Konečny, H. B. McMahan, and A. Talwalkar, "Expanding the reach of federated learning by reducing client resource requirements," *arXiv preprint arXiv:1812.07210*, 2018.
71. Z. Tao and Q. Li, "esgd: Communication efficient distributed deep learning on the edge," in *{USENIX} Workshop on Hot Topics in Edge Computing (HotEdge 18)*, 2018.
72. S. Hochreiter and J. Schmidhuber, "Long short-term memory," *Neural computation*, vol. 9, no. 8, pp. 1735–1780, 1997.
73. Y. Liu, Y. Kang, X. Zhang, L. Li, Y. Cheng, T. Chen, M. Hong, and Q. Yang, "A communication efficient vertical federated learning framework," *arXiv preprint arXiv:1912.11187*, 2019.
74. X. Yao, C. Huang, and L. Sun, "Two-stream federated learning: Reduce the communication costs," in *2018 IEEE Visual Communications and Image Processing (VCIP)*. IEEE, 2018, pp. 1–4.
75. M. Long, Y. Cao, J. Wang, and M. I. Jordan, "Learning transferable features with deep adaptation networks," in *Proceedings of the 32nd International Conference on International Conference on Machine Learning-Volume 37*. JMLR.org, 2015, pp. 97–105.
76. M. Long, J. Wang, G. Ding, J. Sun, and P. S. Yu, "Transfer feature learning with joint distribution adaptation," in *Proceedings of the IEEE international conference on computer vision*, 2013, pp. 2200–2207.
77. A. Krizhevsky, I. Sutskever, and G. E. Hinton, "Imagenet classification with deep convolutional neural networks," in *Advances in neural information processing systems*, 2012, pp. 1097–1105.
78. L. Liu, J. Zhang, S. Song, and K. B. Letaief, "Edge-assisted hierarchical federated learning with non-iid data," *arXiv preprint arXiv:1905.06641*, 2019.
79. G. Cohen, S. Afshar, J. Tapson, and A. van Schaik, "Emnist: an extension of mnist to handwritten letters," *arXiv preprint arXiv:1702.05373*, 2017.

80. N. Strom, "Scalable distributed DNN training using commodity GPU cloud computing," in *Sixteenth Annual Conference of the International Speech Communication Association*, 2015.
81. J. Chen, X. Pan, R. Monga, S. Bengio, and R. Jozefowicz, "Revisiting distributed synchronous sgd," *arXiv preprint arXiv:1604.00981*, 2016.
82. Y. Lin, S. Han, H. Mao, Y. Wang, and W. J. Dally, "Deep gradient compression: Reducing the communication bandwidth for distributed training," *arXiv preprint arXiv:1712.01887*, 2017.
83. K. Hsieh, A. Harlap, N. Vijaykumar, D. Konomis, G. R. Ganger, P. B. Gibbons, and O. Mutlu, "Gaia: Geo-distributed machine learning approaching {LAN} speeds," in *14th {USENIX} Symposium on Networked Systems Design and Implementation ({NSDI} 17)*, 2017, pp. 629–647.
84. D. Anguita, A. Ghio, L. Oneto, X. Parra, and J. L. Reyes-Ortiz, "A public domain dataset for human activity recognition using smartphones." in *Esann*, 2013.
85. M. Buscema and S. Terzi, "Semeion handwritten digit data set," *Center for Machine Learning and Intelligent Systems, California, USA*, 2009.
86. M. R. Sprague, A. Jalalirad, M. Scavuzzo, C. Capota, M. Neun, L. Do, and M. Kopp, "Asynchronous federated learning for geospatial applications," in *Joint European Conference on Machine Learning and Knowledge Discovery in Databases*. Springer, 2018, pp. 21–28.
87. M. I. Jordan, J. D. Lee, and Y. Yang, "Communication-efficient distributed statistical inference," *Journal of the American Statistical Association*, vol. 114, no. 526, pp. 668–681, 2019.
88. T. Nishio and R. Yonetani, "Client selection for federated learning with heterogeneous resources in mobile edge," in *ICC 2019-2019 IEEE International Conference on Communications (ICC)*. IEEE, 2019, pp. 1–7.
89. M. Sviridenko, "A note on maximizing a submodular set function subject to a knapsack constraint," *Operations Research Letters*, vol. 32, no. 1, pp. 41–43, 2004.
90. M. Mohri, G. Sivek, and A. T. Suresh, "Agnostic federated learning," in *International Conference on Machine Learning*. PMLR, 2019, pp. 4615–4625.
91. N. Yoshida, T. Nishio, M. Morikura, K. Yamamoto, and R. Yonetani, "Hybrid-fl: Cooperative learning mechanism using non-iid data in wireless networks," *arXiv preprint arXiv:1905.07210*, 2019.
92. T. T. Anh, N. C. Luong, D. Niyato, D. I. Kim, and L.-C. Wang, "Efficient training management for mobile crowd-machine learning: A deep reinforcement learning approach," *arXiv preprint arXiv:1812.03633*, 2018.
93. H. Van Hasselt, A. Guez, and D. Silver, "Deep reinforcement learning with double q-learning," in *Proceedings of the AAAI conference on artificial intelligence*, vol. 30, no. 1, 2016.
94. M. J. Neely, E. Modiano, and C.-P. Li, "Fairness and optimal stochastic control for heterogeneous networks," *IEEE/ACM Transactions On Networking*, vol. 16, no. 2, pp. 396–409, 2008.
95. S. Corbett-Davies and S. Goel, "The measure and mismeasure of fairness: A critical review of fair machine learning," *arXiv preprint arXiv:1808.00023*, 2018.
96. T. Li, M. Sanjabi, and V. Smith, "Fair resource allocation in federated learning," *arXiv preprint arXiv:1905.10497*, 2019.
97. M. Chen, H. V. Poor, W. Saad, and S. Cui, "Convergence time optimization for federated learning over wireless networks," *arXiv preprint arXiv:2001.07845*, 2020.
98. D. López-Pérez, A. Valcarce, G. De La Roche, and J. Zhang, "Ofdma femtocells: A roadmap on interference avoidance," *IEEE Communications Magazine*, vol. 47, no. 9, pp. 41–48, 2009.
99. G. Zhu, Y. Wang, and K. Huang, "Low-latency broadband analog aggregation for federated edge learning," *arXiv preprint arXiv:1812.11494*, 2018.
100. B. Nazer and M. Gastpar, "Compute-and-forward: Harnessing interference through structured codes," *IEEE Transactions on Information Theory*, vol. 57, no. 10, pp. 6463–6486, 2011.
101. M. M. Amiri and D. Gunduz, "Federated learning over wireless fading channels," *arXiv preprint arXiv:1907.09769*, 2019.

102. K. Yang, T. Jiang, Y. Shi, and Z. Ding, "Federated learning via over-the-air computation," *arXiv preprint arXiv:1812.11750*, 2018.
103. N. S. Keskar, D. Mudigere, J. Nocedal, M. Smelyanskiy, and P. T. P. Tang, "On large-batch training for deep learning: Generalization gap and sharp minima," *arXiv preprint arXiv:1609.04836*, 2016.
104. S. Boyd and L. Vandenberghe, *Convex optimization*. Cambridge University Press, 2004.
105. P. D. Tao *et al.*, "The DC (difference of convex functions) programming and DCA revisited with DC models of real world nonconvex optimization problems," *Annals of operations research*, vol. 133, no. 1-4, pp. 23–46, 2005.
106. Z.-Q. Luo, N. D. Sidiropoulos, P. Tseng, and S. Zhang, "Approximation bounds for quadratic optimization with homogeneous quadratic constraints," *SIAM Journal on optimization*, vol. 18, no. 1, pp. 1–28, 2007.
107. C. Xie, S. Koyejo, and I. Gupta, "Asynchronous federated optimization," *arXiv preprint arXiv:1903.03934*, 2019.
108. K. Bonawitz, H. Eichner, W. Grieskamp, D. Huba, A. Ingerman, V. Ivanov, C. Kiddon, J. Konecny, S. Mazzocchi, H. B. McMahan *et al.*, "Towards federated learning at scale: System design," *arXiv preprint arXiv:1902.01046*, 2019.
109. S. Feng, D. Niyato, P. Wang, D. I. Kim, and Y.-C. Liang, "Joint service pricing and cooperative relay communication for federated learning," *arXiv preprint arXiv:1811.12082*, 2018.
110. M. J. Osborne *et al.*, *An introduction to game theory*. Oxford University Press New York, 2004, vol. 3, no. 3.
111. Y. Sarikaya and O. Ercetin, "Motivating workers in federated learning: A stackelberg game perspective," *arXiv preprint arXiv:1908.03092*, 2019.
112. L. U. Khan, N. H. Tran, S. R. Pandey, W. Saad, Z. Han, M. N. Nguyen, and C. S. Hong, "Federated learning for edge networks: Resource optimization and incentive mechanism," *arXiv preprint arXiv:1911.05642*, 2019.
113. Y. Zhan, P. Li, Z. Qu, D. Zeng, and S. Guo, "A learning-based incentive mechanism for federated learning," *IEEE Internet of Things Journal*, 2020.
114. J. Kang, Z. Xiong, D. Niyato, H. Yu, Y.-C. Liang, and D. I. Kim, "Incentive design for efficient federated learning in mobile networks: A contract theory approach," *arXiv preprint arXiv:1905.07479*, 2019.
115. P. Bolton, M. Dewatripont *et al.*, *Contract theory*. MIT Press, 2005.
116. R. Jurca and B. Faltings, "An incentive compatible reputation mechanism," in *EEE International Conference on E-Commerce, 2003. CEC 2003*. IEEE, 2003, pp. 285–292.
117. R. Dennis and G. Owen, "Rep on the block: A next generation reputation system based on the blockchain," in *2015 10th International Conference for Internet Technology and Secured Transactions (ICITST)*. IEEE, 2015, pp. 131–138.
118. L. Melis, C. Song, E. De Cristofaro, and V. Shmatikov, "Exploiting unintended feature leakage in collaborative learning." IEEE, 2019.
119. H.-W. Ng and S. Winkler, "A data-driven approach to cleaning large face datasets," in *2014 IEEE International Conference on Image Processing (ICIP)*. IEEE, 2014, pp. 343–347.
120. H. T. Nguyen, N. C. Luong, J. Zhao, C. Yuen, and D. Niyato, "Resource allocation in mobility-aware federated learning networks: A deep reinforcement learning approach," *arXiv preprint arXiv:1910.09172*, 2019.
121. G. Ateniese, L. V. Mancini, A. Spognardi, A. Villani, D. Vitali, and G. Felici, "Hacking smart machines with smarter ones: How to extract meaningful data from machine learning classifiers," *International Journal of Security*, vol. 10, no. 3, pp. 137–150, 2015.
122. M. Fredrikson, S. Jha, and T. Ristenpart, "Model inversion attacks that exploit confidence information and basic countermeasures," in *Proceedings of the 22nd ACM SIGSAC Conference on Computer and Communications Security*. ACM, 2015, pp. 1322–1333.
123. F. Tramèr, F. Zhang, A. Juels, M. K. Reiter, and T. Ristenpart, "Stealing machine learning models via prediction APIs," in *25th {USENIX} Security Symposium ({USENIX} Security 16)*, 2016, pp. 601–618.

124. R. Shokri, M. Stronati, C. Song, and V. Shmatikov, "Membership inference attacks against machine learning models," in *2017 IEEE Symposium on Security and Privacy (SP)*. IEEE, 2017, pp. 3–18.
125. N. Papernot, P. McDaniel, and I. Goodfellow, "Transferability in machine learning: from phenomena to black-box attacks using adversarial samples," *arXiv preprint arXiv:1605.07277*, 2016.
126. P. Laskov *et al.*, "Practical evasion of a learning-based classifier: A case study," in *2014 IEEE symposium on security and privacy*. IEEE, 2014, pp. 197–211.
127. C. Dwork, F. McSherry, K. Nissim, and A. Smith, "Calibrating noise to sensitivity in private data analysis," in *Theory of cryptography conference*. Springer, 2006, pp. 265–284.
128. R. C. Geyer, T. Klein, and M. Nabi, "Differentially private federated learning: A client level perspective," *arXiv preprint arXiv:1712.07557*, 2017.
129. R. Shokri and V. Shmatikov, "Privacy-preserving deep learning," in *Proceedings of the 22nd ACM SIGSAC conference on computer and communications security*. ACM, 2015, pp. 1310–1321.
130. B. Hitaj, G. Ateniese, and F. Pérez-Cruz, "Deep models under the GAN: information leakage from collaborative deep learning," in *Proceedings of the 2017 ACM SIGSAC Conference on Computer and Communications Security*. ACM, 2017, pp. 603–618.
131. I. Goodfellow, J. Pouget-Abadie, M. Mirza, B. Xu, D. Warde-Farley, S. Ozair, A. Courville, and Y. Bengio, "Generative adversarial nets," in *Advances in neural information processing systems*, 2014, pp. 2672–2680.
132. Y. Liu, Z. Ma, S. Ma, S. Nepal, and R. Deng, "Boosting privately: Privacy-preserving federated extreme boosting for mobile crowdsensing," *https://arxiv.org/pdf/1907.10218.pdf*, 2019.
133. A. Triastcyn and B. Faltings, "Federated generative privacy," *arXiv preprint arXiv:1910.08385*, 2019.
134. Y. Aono, T. Hayashi, L. Wang, S. Moriai *et al.*, "Privacy-preserving deep learning via additively homomorphic encryption," *IEEE Transactions on Information Forensics and Security*, vol. 13, no. 5, pp. 1333–1345, 2017.
135. K. Bonawitz, V. Ivanov, B. Kreuter, A. Marcedone, H. B. McMahan, S. Patel, D. Ramage, A. Segal, and K. Seth, "Practical secure aggregation for privacy-preserving machine learning," in *Proceedings of the 2017 ACM SIGSAC Conference on Computer and Communications Security*. ACM, 2017, pp. 1175–1191.
136. M. Hao, H. Li, G. Xu, S. Liu, and H. Yang, "Towards efficient and privacy-preserving federated deep learning," in *2019 IEEE International Conference on Communications*. IEEE, 2019, pp. 1–6.
137. X. Chen, C. Liu, B. Li, K. Lu, and D. Song, "Targeted backdoor attacks on deep learning systems using data poisoning," *arXiv preprint arXiv:1712.05526*, 2017.
138. C. Fung, C. J. Yoon, and I. Beschastnikh, "Mitigating sybils in federated learning poisoning," *arXiv preprint arXiv:1808.04866*, 2018.
139. A. Asuncion and D. Newman, "UCI machine learning repository," 2007.
140. A. N. Bhagoji, S. Chakraborty, P. Mittal, and S. Calo, "Analyzing federated learning through an adversarial lens," *arXiv preprint arXiv:1811.12470*, 2018.
141. E. Bagdasaryan, A. Veit, Y. Hua, D. Estrin, and V. Shmatikov, "How to backdoor federated learning," *arXiv preprint arXiv:1807.00459*, 2018.
142. H. Kim, J. Park, M. Bennis, and S.-L. Kim, "On-device federated learning via blockchain and its latency analysis," *arXiv preprint arXiv:1808.03949*, 2018.
143. J. Weng, J. Weng, J. Zhang, M. Li, Y. Zhang, and W. Luo, "Deepchain: Auditable and privacy-preserving deep learning with blockchain-based incentive," *Cryptology ePrint Archive, Report 2018/679*, 2018.
144. I. Stojmenovic, S. Wen, X. Huang, and H. Luan, "An overview of fog computing and its security issues," *Concurrency and Computation: Practice and Experience*, vol. 28, no. 10, pp. 2991–3005, 2016.

145. M. Gerla, E.-K. Lee, G. Pau, and U. Lee, "Internet of vehicles: From intelligent grid to autonomous cars and vehicular clouds," in *2014 IEEE world forum on internet of things (WF-IoT)*. IEEE, 2014, pp. 241–246.
146. K. K. Nguyen, D. T. Hoang, D. Niyato, P. Wang, D. Nguyen, and E. Dutkiewicz, "Cyberattack detection in mobile cloud computing: A deep learning approach," in *2018 IEEE Wireless Communications and Networking Conference (WCNC)*. IEEE, 2018, pp. 1–6.
147. T. U. of New Brunswick, "Nsl-kdd https://www.unb.ca/cic/datasets/nsl.html."
148. N. Moustafa and J. Slay, "Unsw-nb15: a comprehensive data set for network intrusion detection systems (unsw-nb15 network data set)," in *2015 military communications and information systems conference (MilCIS)*. IEEE, 2015, pp. 1–6.
149. A. Abeshu and N. Chilamkurti, "Deep learning: the frontier for distributed attack detection in fog-to-things computing," *IEEE Communications Magazine*, vol. 56, no. 2, pp. 169–175, 2018.
150. T. D. Nguyen, S. Marchal, M. Miettinen, H. Fereidooni, N. Asokan, and A.-R. Sadeghi, "Dïot: A federated self-learning anomaly detection system for IoT," in *2019 IEEE 39th International Conference on Distributed Computing Systems (ICDCS)*. IEEE, 2019, pp. 756–767.
151. D. Preuveneers, V. Rimmer, I. Tsingenopoulos, J. Spooren, W. Joosen, and E. Ilie-Zudor, "Chained anomaly detection models for federated learning: An intrusion detection case study," *Applied Sciences*, vol. 8, no. 12, p. 2663, 2018.
152. J. Ren, H. Wang, T. Hou, S. Zheng, and C. Tang, "Federated learning-based computation offloading optimization in edge computing-supported internet of things," *IEEE Access*, vol. 7, pp. 69 194–69 201, 2019.
153. Z. Yu, J. Hu, G. Min, H. Lu, Z. Zhao, H. Wang, and N. Georgalas, "Federated learning based proactive content caching in edge computing," in *2018 IEEE Global Communications Conference (GLOBECOM)*. IEEE, 2018, pp. 1–6.
154. Y. Qian, L. Hu, J. Chen, X. Guan, M. M. Hassan, and A. Alelaiwi, "Privacy-aware service placement for mobile edge computing via federated learning," *Information Sciences*, vol. 505, pp. 562–570, 2019.
155. M. Chen, O. Semiari, W. Saad, X. Liu, and C. Yin, "Federated echo state learning for minimizing breaks in presence in wireless virtual reality networks," *IEEE Transactions on Wireless Communications*, vol. 19, no. 1, pp. 177–191, 2019.
156. Y. Saputra, H. Dinh, D. Nguyen, E. Dutkiewicz, M. Mueck, and S. Srikanteswara, in *2019 IEEE Global Communications Conference (GLOBECOM)*. IEEE, 2019, pp. 1–6.
157. S. J. Pan and Q. Yang, "A survey on transfer learning," *IEEE Transactions on knowledge and data engineering*, vol. 22, no. 10, pp. 1345–1359, 2009.
158. S. Priya and D. J. Inman, *Energy harvesting technologies*. Springer, 2009, vol. 21.
159. O. Chapelle and L. Li, "An empirical evaluation of Thompson sampling," in *Advances in neural information processing systems*, 2011, pp. 2249–2257.
160. J. Chung, H.-J. Yoon, and H. J. Gardner, "Analysis of break in presence during game play using a linear mixed model," *ETRI journal*, vol. 32, no. 5, pp. 687–694, 2010.
161. M. Chen, M. Mozaffari, W. Saad, C. Yin, M. Debbah, and C. S. Hong, "Caching in the sky: Proactive deployment of cache-enabled unmanned aerial vehicles for optimized quality-of-experience," *IEEE Journal on Selected Areas in Communications*, vol. 35, no. 5, pp. 1046–1061, 2017.
162. O. Guéant, J.-M. Lasry, and P.-L. Lions, "Mean field games and applications," in *Paris-Princeton lectures on mathematical finance 2010*. Springer, 2011, pp. 205–266.
163. S. Samarakoon, M. Bennis, W. Saad, M. Debbah, "Federated learning for ultra-reliable low-latency V2V communications," in *2018 IEEE Global Communications Conference (GLOBECOM)*. IEEE, 2018, pp. 1–7.
164. L. De Haan and A. Ferreira, *Extreme value theory: an introduction*. Springer Science & Business Media, 2007.
165. M. J. Neely, "Stochastic network optimization with application to communication and queueing systems," *Synthesis Lectures on Communication Networks*, vol. 3, no. 1, pp. 1–211, 2010.

166. D. Ye, R. Yu, M. Pan, and Z. Han, “Federated learning in vehicular edge computing: A selective model aggregation approach,” *IEEE Access*, 2020.
167. P. You and Z. Yang, “Efficient optimal scheduling of charging station with multiple electric vehicles via v2v,” in *2014 IEEE International Conference on Smart Grid Communications (SmartGridComm)*. IEEE, 2014, pp. 716–721.
168. P. Bradley, K. Bennett, and A. Demiriz, “Constrained k-means clustering,” *Microsoft Research, Redmond*, vol. 20, no. 0, p. 0, 2000.
169. W. Li, T. Logenthiran, V.-T. Phan, and W. L. Woo, “Implemented IoT-based self-learning home management system (SHMS) for Singapore,” *IEEE Internet of Things Journal*, vol. 5, no. 3, pp. 2212–2219, 2018.
170. R. Boutaba, M. A. Salahuddin, N. Limam, S. Ayoubi, N. Shahriar, F. Estrada-Solano, and O. M. Caicedo, “A comprehensive survey on machine learning for networking: evolution, applications and research opportunities,” *Journal of Internet Services and Applications*, vol. 9, no. 1, p. 16, 2018.
171. Z. Yang, M. Chen, W. Saad, C. S. Hong, and M. Shikh-Bahaei, “Energy efficient federated learning over wireless communication networks,” *IEEE Transactions on Wireless Communications*, vol. 20, no. 3, pp. 1935–1949, 2020.
172. F. Li and Y. Wang, “Routing in vehicular ad hoc networks: A survey,” *IEEE Vehicular Technology Magazine*, vol. 2, no. 2, pp. 12–22, 2007.
173. H. Hartenstein and L. Laberteaux, “A tutorial survey on vehicular ad hoc networks,” *IEEE Communications Magazine*, vol. 46, no. 6, pp. 164–171, 2008.
174. W. Xu, H. Zhou, N. Cheng, F. Lyu, W. Shi, J. Chen, and X. Shen, “Internet of vehicles in big data era,” *IEEE/CAA Journal of Automatica Sinica*, vol. 5, no. 1, pp. 19–35, 2017.
175. J. Wan, J. Liu, Z. Shao, A. V. Vasilakos, M. Imran, and K. Zhou, “Mobile crowd sensing for traffic prediction in internet of vehicles,” *Sensors*, vol. 16, no. 1, p. 88, 2016.
176. F. Yang, S. Wang, J. Li, Z. Liu, and Q. Sun, “An overview of internet of vehicles,” *China communications*, vol. 11, no. 10, pp. 1–15, 2014.
177. J. Wang, C. Jiang, Z. Han, Y. Ren, and L. Hanzo, “Internet of vehicles: Sensing-aided transportation information collection and diffusion,” *IEEE Transactions on Vehicular Technology*, vol. 67, no. 5, pp. 3813–3825, 2018.
178. P. M. Kumar, G. Manogaran, R. Sundarasekar, N. Chilamkurti, R. Varatharajan *et al.*, “Ant colony optimization algorithm with internet of vehicles for intelligent traffic control system,” *Computer Networks*, vol. 144, pp. 154–162, 2018.
179. M. Florian, S. Finster, and I. Baumgart, “Privacy-preserving cooperative route planning,” *IEEE Internet of Things Journal*, vol. 1, no. 6, pp. 590–599, 2014.
180. L.-M. Ang, K. P. Seng, G. K. Ijemaru, and A. M. Zungeru, “Deployment of IoV for smart cities: applications, architecture, and challenges,” *IEEE Access*, vol. 7, pp. 6473–6492, 2018.
181. H. Zhou, H. Kong, L. Wei, D. Creighton, and S. Nahavandi, “Efficient road detection and tracking for unmanned aerial vehicle,” *IEEE Transactions on Intelligent Transportation Systems*, vol. 16, no. 1, pp. 297–309, 2014.
182. H. Zhou, L. Wei, M. Fielding, D. Creighton, S. Deshpande, and S. Nahavandi, “Car park occupancy analysis using UAV images,” in *2017 IEEE International Conference on Systems, Man, and Cybernetics (SMC)*. IEEE, 2017, pp. 3261–3265.
183. M. Elloumi, R. Dhaou, B. Escrig, H. Idoudi, and L. A. Saidane, “Monitoring road traffic with a UAV-based system,” in *2018 IEEE Wireless Communications and Networking Conference (WCNC)*. IEEE, 2018, pp. 1–6.
184. B. Coifman, M. McCord, R. G. Mishalani, M. Iswalt, and Y. Ji, “Roadway traffic monitoring from an unmanned aerial vehicle,” in *IEE Proceedings-Intelligent Transport Systems*, vol. 153, no. 1. IET, 2006, pp. 11–20.
185. R. Ke, Z. Li, J. Tang, Z. Pan, and Y. Wang, “Real-time traffic flow parameter estimation from UAV video based on ensemble classifier and optical flow,” *IEEE Transactions on Intelligent Transportation Systems*, vol. 20, no. 1, pp. 54–64, 2018.
186. H. Binol, E. Bulut, K. Akkaya, and I. Guvenc, “Time optimal multi-UAV path planning for gathering its data from roadside units,” in *2018 IEEE 88th Vehicular Technology Conference (VTC-Fall)*. IEEE, 2018, pp. 1–5.

187. L. Zhang, Z. Zhao, Q. Wu, H. Zhao, H. Xu, and X. Wu, "Energy-aware dynamic resource allocation in UAV assisted mobile edge computing over social internet of vehicles," *IEEE Access*, vol. 6, pp. 56 700–56 715, 2018.
188. O. Bekkouche, T. Taleb, and M. Bagaa, "UAVs traffic control based on multi-access edge computing," in *2018 IEEE Global Communications Conference (GLOBECOM)*. IEEE, 2018, pp. 1–6.
189. M. Gharibi, R. Boutaba, and S. L. Waslander, "Internet of drones," *IEEE Access*, vol. 4, pp. 1148–1162, 2016.
190. A. Koubâa and B. Qureshi, "Dronetrack: Cloud-based real-time object tracking using unmanned aerial vehicles over the internet," *IEEE Access*, vol. 6, pp. 13 810–13 824, 2018.
191. A. Koubâa, B. Qureshi, M.-F. Sriti, Y. Javed, and E. Tovar, "A service-oriented cloud-based management system for the internet-of-drones," in *2017 IEEE International Conference on Autonomous Robot Systems and Competitions (ICARSC)*. IEEE, 2017, pp. 329–335.
192. L. Gupta, R. Jain, and G. Vaszkun, "Survey of important issues in UAV communication networks," *IEEE Communications Surveys & Tutorials*, vol. 18, no. 2, pp. 1123–1152, 2015.
193. Y. Zeng and R. Zhang, "Energy-efficient UAV communication with trajectory optimization," *IEEE Transactions on Wireless Communications*, vol. 16, no. 6, pp. 3747–3760, 2017.
194. L. E. Dubins and D. A. Freedman, "Machiavelli and the Gale-Shapley algorithm," *The American Mathematical Monthly*, vol. 88, no. 7, pp. 485–494, 1981.
195. Z. Zhou, J. Feng, B. Gu, B. Ai, S. Mumtaz, J. Rodriguez, and M. Guizani, "When mobile crowd sensing meets UAV: Energy-efficient task assignment and route planning," *IEEE Transactions on Communications*, vol. 66, no. 11, pp. 5526–5538, 2018.
196. G. Karypis and V. Kumar, "Multilevel graph partitioning schemes," in *ICPP (3)*, 1995, pp. 113–122.
197. Y. Zeng, J. Xu, and R. Zhang, "Energy minimization for wireless communication with rotary-wing UAV," *IEEE Transactions on Wireless Communications*, vol. 18, no. 4, pp. 2329–2345, 2019.
198. Q. Zhang, W. Saad, M. Bennis, X. Lu, M. Debbah, and W. Zuo, "Predictive deployment of UAV base stations in wireless networks: Machine learning meets contract theory," *arXiv preprint arXiv:1811.01149*, 2018.
199. Y. Mao, J. Zhang, and K. B. Letaief, "Dynamic computation offloading for mobile-edge computing with energy harvesting devices," *IEEE Journal on Selected Areas in Communications*, vol. 34, no. 12, pp. 3590–3605, 2016.
200. Q. Liu, S. Huang, J. Opadere, and T. Han, "An edge network orchestrator for mobile augmented reality," in *IEEE INFOCOM 2018-IEEE Conference on Computer Communications*. IEEE, 2018, pp. 756–764.
201. Z. Wang, L. Gao, and J. Huang, "Multi-cap optimization for wireless data plans with time flexibility," *IEEE Transactions on Mobile Computing*, vol. 19, no. 9, pp. 2145–2159, 2019.
202. L. Gao, X. Wang, Y. Xu, and Q. Zhang, "Spectrum trading in cognitive radio networks: A contract-theoretic modeling approach," *IEEE Journal on Selected Areas in Communications*, vol. 29, no. 4, pp. 843–855, 2011.
203. A. E. Roth and M. Sotomayor, "Two-sided matching," *Handbook of game theory with economic applications*, vol. 1, pp. 485–541, 1992.
204. Z. Zhou, P. Liu, J. Feng, Y. Zhang, S. Mumtaz, and J. Rodriguez, "Computation resource allocation and task assignment optimization in vehicular fog computing: A contract-matching approach," *IEEE Transactions on Vehicular Technology*, vol. 68, no. 4, pp. 3113–3125, 2019.
205. G. O'Malley, "Algorithmic aspects of stable matching problems," Ph.D. dissertation, University of Glasgow, 2007.
206. S. Sawadsitang, D. Niyato, P.-S. Tan, and P. Wang, "Joint ground and aerial package delivery services: A stochastic optimization approach," *IEEE Transactions on Intelligent Transportation Systems*, vol. 20, no. 6, pp. 2241–2254, 2018.
207. Y. Guo, S. Gu, Q. Zhang, N. Zhang, and W. Xiang, "A coded distributed computing framework for task offloading from multi-UAV to edge servers," in *2021 IEEE Wireless Communications and Networking Conference (WCNC)*. IEEE, 2021, pp. 1–6.

208. S. Prakash, S. Dhakal, M. R. Akdeniz, Y. Yona, S. Talwar, S. Avestimehr, and N. Himayat, "Coded computing for low-latency federated learning over wireless edge networks," *IEEE Journal on Selected Areas in Communications*, vol. 39, no. 1, pp. 233–250, 2020.
209. (2019, Aug. 5) Internet of Things Ecosystem and Trend. IDC. https://www.idc.com/getdoc.jsp?containerId=IDC_P24793.
210. W. Zhuang, Q. Ye, F. Lyu, N. Cheng, and J. Ren, "SDN/NFV-Empowered Future IoV With Enhanced Communication, Computing, and Caching," *Proceedings of the IEEE*, vol. 108, no. 2, pp. 274–291, 2020.
211. Q. Yang, Y. Liu, T. Chen, and Y. Tong, "Federated Machine Learning: Concept and Applications," *ACM Trans. Intell. Syst. Technol.*, vol. 10, no. 2, Jan. 2019. [Online]. Available: https://doi.org.remotexs.ntu.edu.sg/10.1145/3298981
212. Y. Hu, J. Lam, and J. Liang, "Consensus Control of Multi-Agent Systems with Missing Data in Actuators and Markovian Communication Failure," *International Journal of Systems Science*, vol. 44, no. 10, pp. 1867–1878, 2013. [Online]. Available: https://doi.org/10.1080/00207721.2012.670298
213. S. Abdullah and K. Yang, "An Energy Efficient Message Scheduling Algorithm Considering Node Failure in IoT Environment," *Wireless Personal Communications*, vol. 79, no. 3, pp. 1815–1835, 2014. [Online]. Available: https://doi.org/10.1007/s11277-014-1960-3
214. H. Li, K. Ota, and M. Dong, "Learning IoT in edge: Deep learning for the internet of things with edge computing," *IEEE Network*, vol. 32, no. 1, pp. 96–101, 2018.
215. P. Zhan, K. Yu, and A. L. Swindlehurst, "Wireless Relay Communications with Unmanned Aerial Vehicles: Performance and Optimization," *IEEE Transactions on Aerospace and Electronic Systems*, vol. 47, no. 3, pp. 2068–2085, 2011.
216. P. Basu, J. Redi, and V. Shurbanov, "Coordinated Flocking of UAVs for Improved Connectivity of Mobile Ground Nodes," in *IEEE MILCOM 2004. Military Communications Conference, 2004*, vol. 3, 2004, pp. 1628–1634.
217. Y. Liu, T. Chen, and Q. Yang, "Secure federated transfer learning," *arXiv preprint arXiv:1812.03337*, 2018.
218. M. Kantarcioglu and R. Nix, "Incentive Compatible Distributed Data Mining," in *2010 IEEE Second International Conference on Social Computing*, 2010, pp. 735–742.
219. S. Wang, T. Tuor, T. Salonidis, K. K. Leung, C. Makaya, T. He, and K. Chan, "Adaptive Federated Learning in Resource Constrained Edge Computing Systems," *IEEE Journal on Selected Areas in Communications*, vol. 37, no. 6, pp. 1205–1221, June 2019.
220. N. Nguyen and M. M. H. Khan, "Context Aware Data Acquisition Framework for Dynamic Data Driven Applications Systems (DDDAS)," in *MILCOM 2013 - 2013 IEEE Military Communications Conference*, Nov 2013, pp. 334–341.
221. S. Xie and Z. Chen, "Anomaly Detection and Redundancy Elimination of Big Sensor Data in Internet of Things," *arXiv preprint arXiv:1703.03225*, 2017.
222. S.-J. Yoo, J. hyun Park, S. hee Kim, and A. Shrestha, "Flying Path Optimization in UAV-assisted IoT sensor networks," *ICT Express*, vol. 2, no. 3, pp. 140–144, 2016, special Issue on ICT Convergence in the Internet of Things (IoT). [Online]. Available: http://www.sciencedirect.com/science/article/pii/S2405959516301096
223. J. Zhang, X. Hu, Z. Ning, E. C. H. Ngai, L. Zhou, J. Wei, J. Cheng, and B. Hu, "Energy-Latency Tradeoff for Energy-Aware Offloading in Mobile Edge Computing Networks," *IEEE Internet of Things Journal*, vol. 5, no. 4, pp. 2633–2645, Aug 2018.
224. A. P. Miettinen and J. K. Nurminen, "Energy Efficiency of Mobile Clients in Cloud Computing," *HotCloud*, vol. 10, no. 4-4, p. 19, 2010.
225. M. Chen, Z. Yang, W. Saad, C. Yin, H. V. Poor, and S. Cui, "A Joint Learning and Communications Framework for Federated Learning Over Wireless Networks," *arXiv preprint arXiv:1909.07972*, 2019.
226. F. Z. Kaddour, E. Vivier, L. Mroueh, M. Pischella, and P. Martins, "Green Opportunistic and Efficient Resource Block Allocation Algorithm for LTE Uplink Networks," *IEEE Transactions on Vehicular Technology*, vol. 64, no. 10, pp. 4537–4550, 2015.

227. N. Guan, Y. Zhou, L. Tian, G. Sun, and J. Shi, "QoS Guaranteed Resource Block Allocation Algorithm for LTE systems," in *2011 IEEE 7th International Conference on Wireless and Mobile Computing, Networking and Communications (WiMob)*, 2011, pp. 307–312.
228. J. Lu, S. Wan, X. Chen, and P. Fan, "Energy-Efficient 3D UAV-BS Placement versus Mobile Users' Density and Circuit Power," in *2017 IEEE Globecom Workshops (GC Wkshps)*, 2017, pp. 1–6.
229. W. Saad, Z. Han, M. Debbah, A. Hjorungnes, and T. Basar, "Coalitional Game Theory for Communication Networks," *IEEE Signal Processing Magazine*, vol. 26, no. 5, pp. 77–97, Sep. 2009.
230. W. Saad, Z. Han, M. Debbah, and A. Hjorungnes, "A Distributed Merge and Split Algorithm for Fair Cooperation in Wireless Networks," in *ICC Workshops - 2008 IEEE International Conference on Communications Workshops*, 2008, pp. 311–315.
231. K. R. Apt and T. Radzik, "Stable Partitions in Coalitional Games," *arXiv preprint cs/0605132*, 2006.
232. A. Bogomolnaia and M. O. Jackson, "The Stability of Hedonic Coalition Structures," *Games and Economic Behavior*, vol. 38, no. 2, pp. 201–230, 2002. [Online]. Available: https://www.sciencedirect.com/science/article/pii/S0899825601908772
233. L. Lu, J. Yu, Y. Zhu, and M. Li, "A Double Auction Mechanism to Bridge Users' Task Requirements and Providers' Resources in Two-Sided Cloud Markets," *IEEE Transactions on Parallel and Distributed Systems*, vol. 29, no. 4, pp. 720–733, 2018.
234. J. Baek, S. I. Han, and Y. Han, "Optimal Resource Allocation for Non-Orthogonal Transmission in UAV Relay Systems," *IEEE Wireless Communications Letters*, vol. 7, no. 3, pp. 356–359, 2018.
235. W. Y. B. Lim, N. C. Luong, D. T. Hoang, Y. Jiao, Y.-C. Liang, Q. Yang, D. Niyato, and C. Miao, "Federated learning in mobile edge networks: A comprehensive survey," *IEEE Communications Surveys & Tutorials*, vol. 22, no. 3, pp. 2031–2063, 2020.
236. D. Xu, T. Li, Y. Li, X. Su, S. Tarkoma, and P. Hui, "A survey on edge intelligence," *arXiv preprint arXiv:2003.12172*, 2020.
237. W. Y. B. Lim, J. S. Ng, Z. Xiong, D. Niyato, C. Leung, C. Miao, and Q. Yang, "Incentive mechanism design for resource sharing in collaborative edge learning," *arXiv preprint arXiv:2006.00511*, 2020.
238. M. S. H. Abad, E. Ozfatura, D. Gunduz, and O. Ercetin, "Hierarchical federated learning across heterogeneous cellular networks," in *ICASSP 2020-2020 IEEE International Conference on Acoustics, Speech and Signal Processing (ICASSP)*. IEEE, 2020, pp. 8866–8870.
239. M. Chen, H. V. Poor, W. Saad, and S. Cui, "Wireless communications for collaborative federated learning," *IEEE Communications Magazine*, vol. 58, no. 12, pp. 48–54, 2020.
240. J. W. Weibull, *Evolutionary game theory*. MIT Press, 1997.
241. E. Peltonen, M. Bennis, M. Capobianco, M. Debbah, A. Ding, F. Gil-Castiñeira, M. Jurmu, T. Karvonen, M. Kelanti, A. Kliks *et al.*, "6g white paper on edge intelligence," *arXiv preprint arXiv:2004.14850*, 2020.
242. M. Xie, H. Yin, H. Wang, F. Xu, W. Chen, and S. Wang, "Learning graph-based poi embedding for location-based recommendation," in *Proceedings of the 25th ACM International on Conference on Information and Knowledge Management*, 2016, pp. 15–24.
243. M. Handy, M. Haase, and D. Timmermann, "Low energy adaptive clustering hierarchy with deterministic cluster-head selection," in *4th international workshop on mobile and wireless communications network*. IEEE, 2002, pp. 368–372.
244. H. Junping, J. Yuhui, and D. Liang, "A time-based cluster-head selection algorithm for leach," in *2008 IEEE Symposium on Computers and Communications*. IEEE, 2008, pp. 1172–1176.
245. R. Ferdous, V. Muthukkumarasamy, and E. Sithirasenan, "Trust-based cluster head selection algorithm for mobile ad hoc networks," in *2011 IEEE 10th International Conference on Trust, Security and Privacy in Computing and Communications*. IEEE, 2011, pp. 589–596.
246. G. Zhang, K. Yang, and H.-H. Chen, "Socially aware cluster formation and radio resource allocation in d2d networks," *IEEE Wireless Communications*, vol. 23, no. 4, pp. 68–73, 2016.

247. K. Zhu, E. Hossain, and D. Niyato, "Pricing, spectrum sharing, and service selection in two-tier small cell networks: A hierarchical dynamic game approach," *IEEE Transactions on Mobile Computing*, vol. 13, no. 8, pp. 1843–1856, 2013.
248. X. Gong, L. Duan, X. Chen, and J. Zhang, "When social network effect meets congestion effect in wireless networks: Data usage equilibrium and optimal pricing," *IEEE Journal on Selected Areas in Communications*, vol. 35, no. 2, pp. 449–462, 2017.
249. W. Y. B. Lim, Z. Xiong, C. Miao, D. Niyato, Q. Yang, C. Leung, and H. V. Poor, "Hierarchical incentive mechanism design for federated machine learning in mobile networks," *IEEE Internet of Things Journal*, vol. 7, no. 10, pp. 9575–9588, 2020.
250. W. Y. B. Lim, J. Huang, Z. Xiong, J. Kang, D. Niyato, X.-S. Hua, C. Leung, and C. Miao, "Towards federated learning in UAV-enabled internet of vehicles: A multi-dimensional contract-matching approach," *IEEE Transactions on Intelligent Transportation Systems*, vol. 22, no. 8, pp. 5140–5154, 2021.
251. D. Niyato and E. Hossain, "Dynamics of network selection in heterogeneous wireless networks: An evolutionary game approach," *IEEE Transactions on Vehicular Technology*, vol. 58, no. 4, 2008.
252. J. Hofbauer and K. Sigmund, "Evolutionary game dynamics," *Bulletin of the American mathematical society*, vol. 40, no. 4, pp. 479–519, 2003.
253. X. Gao, S. Feng, D. Niyato, P. Wang, K. Yang, and Y.-C. Liang, "Dynamic access point and service selection in backscatter-assisted RF-powered cognitive networks," *IEEE Internet of Things Journal*, vol. 6, no. 5, pp. 8270–8283, 2019.
254. J. Engwerda, *LQ dynamic optimization and differential games*. John Wiley & Sons, 2005.
255. S. Sastry, "Lyapunov stability theory," in *Nonlinear Systems*. Springer, 1999, pp. 182–234.
256. Y. Zou, S. Feng, D. Niyato, Y. Jiao, S. Gong, and W. Cheng, "Mobile Device Training Strategies in Federated Learning: An Evolutionary Game Approach," in *2019 International Conference on Internet of Things (iThings) and IEEE Green Computing and Communications (GreenCom) and IEEE Cyber, Physical and Social Computing (CPSCom) and IEEE Smart Data (SmartData)*. IEEE, 2019, pp. 874–879.
257. R. B. Myerson, "Optimal auction design," *Mathematics of operations research*, vol. 6, no. 1, pp. 58–73, 1981.
258. P. Dütting, Z. Feng, H. Narasimhan, D. C. Parkes, and S. S. Ravindranath, "Optimal Auctions through Deep Learning," *arXiv preprint arXiv:1706.03459*, 2017.
259. N. C. Luong, Z. Xiong, P. Wang, and D. Niyato, "Optimal auction for edge computing resource management in mobile blockchain networks: A deep learning approach," in *2018 IEEE International Conference on Communications (ICC)*. IEEE, 2018, pp. 1–6.
260. K. Zhu, D. Niyato, and P. Wang, "Optimal bandwidth allocation with dynamic service selection in heterogeneous wireless networks," in *2010 IEEE Global Telecommunications Conference Globecom 2010*. IEEE, 2010, pp. 1–5.
261. M. Lee, S. Hosseinalipour, C. G. Brinton, G. Yu, and H. Dai, "A fast graph neural network-based method for winner determination in multi-unit combinatorial auctions," *IEEE Transactions on Cloud Computing*, 2020.
262. J. Kang, Z. Xiong, C. Jiang, Y. Liu, S. Guo, Y. Zhang, D. Niyato, C. Leung, and C. Miao, "Scalable and communication-efficient decentralized federated edge learning with multi-blockchain framework," in *International Conference on Blockchain and Trustworthy Systems*. Springer, 2020, pp. 152–165.
263. L. Jiang, R. Tan, X. Lou, and G. Lin, "On lightweight privacy-preserving collaborative learning for internet-of-things objects," in *Proceedings of the International Conference on Internet of Things Design and Implementation*, 2019, pp. 70–81.
264. Z. Gu, H. Jamjoom, D. Su, H. Huang, J. Zhang, T. Ma, D. Pendarakis, and I. Molloy, "Reaching data confidentiality and model accountability on the CalTrain," in *2019 49th Annual IEEE/IFIP International Conference on Dependable Systems and Networks (DSN)*. IEEE, 2019, pp. 336–348.
265. A. Albaseer, B. S. Ciftler, M. Abdallah, and A. Al-Fuqaha, "Exploiting unlabeled data in smart cities using federated learning," *arXiv preprint arXiv:2001.04030*, 2020.

266. P. Vepakomma, O. Gupta, T. Swedish, and R. Raskar, "Split learning for health: Distributed deep learning without sharing raw patient data," *arXiv preprint arXiv:1812.00564*, 2018.
267. A. Singh, P. Vepakomma, O. Gupta, and R. Raskar, "Detailed comparison of communication efficiency of split learning and federated learning," *arXiv preprint arXiv:1909.09145*, 2019.
268. I. I. Eliazar and I. M. Sokolov, "Measuring statistical heterogeneity: The Pietra index," *Physica A: Statistical Mechanics and its Applications*, vol. 389, no. 1, pp. 117–125, 2010.
269. J.-S. Leu, T.-H. Chiang, M.-C. Yu, and K.-W. Su, "Energy efficient clustering scheme for prolonging the lifetime of wireless sensor network with isolated nodes," *IEEE Communications Letters*, vol. 19, no. 2, pp. 259–262, 2014.

Index

A
Auction, v, 50, 81, 83–115, 117–145, 147

C
Coalition formation, v, 83–115
Communication efficiency, v, 12, 29, 50, 54, 83–115, 147, 149
Contract theory, 50, 54

E
Edge intelligence, v, 5, 117–119, 147
Edge networks, v, 1–51, 117, 149
Evolutionary game, v, 119–128, 135–140, 143, 144, 147

F
Federated learning (FL), v, 1–51, 53–81, 83–115, 117–145

G
Game theory, v, 119, 121, 143, 144

I
Incentive mechanism, 21, 23, 28–32, 50, 54–56, 62, 77, 81, 118, 143, 147
Internet of Vehicles (IoV), 40, 53–56, 80, 83, 84, 86, 88, 89, 99, 105–107, 114, 115

M
Mobile edge computing (MEC), 2, 5, 19, 39–49

P
Privacy and security, 4, 12, 23, 32–39, 147, 150

R
Resource allocation, v, 4, 5, 11, 12, 15, 18–32, 39, 50, 51, 118, 119, 143, 147, 148, 150

U
Unmanned aerial vehicles (UAVs), v, 51, 53–81, 83–115, 147

W. Y. B. Lim et al., *Federated Learning Over Wireless Edge Networks*, Wireless Networks, https://doi.org/10.1007/978-3-031-07838-5

The manufacturer's authorised representative in the EU is Springer Nature Customer Service Centre GmbH, Europaplatz 3, 69115 Heidelberg, Germany. If you have any concerns regarding our products, please contact ProductSafety@springernature.com

Printed and bound by CPI Group (UK) Ltd, Croydon, CR0 4YY
15/07/2026
02167645-0012